D1498363

Intimacy, Transcendence, and Psychology

Intimacy, Transcendence, and Psychology

Closeness and Openness in Everyday Life

Steen Halling

palgrave
macmillan

INTIMACY, TRANSCENDENCE, AND PSYCHOLOGY
Copyright © Steen Halling, 2008.

First published in 2008 by
PALGRAVE MACMILLAN™
175 Fifth Avenue, New York, N.Y. 10010 and
Houndmills, Basingstoke, Hampshire, England RG21 6XS.
Companies and representatives throughout the world.

PALGRAVE MACMILLAN is the global academic imprint of the Palgrave Macmillan division of St. Martin's Press, LLC and of Palgrave Macmillan Ltd. Macmillan® is a registered trademark in the United States, United Kingdom and other countries. Palgrave is a registered trademark in the European Union and other countries.

ISBN-13: 978-0-230-60045-4
ISBN-10: 0-230-60045-X

Library of Congress Cataloging-in-Publication Data is available from the Library of Congress.

A catalogue record for this book is available from the British Library.

Design by Macmillan India Ltd.

First edition: January 2008

10 9 8 7 6 5 4 3 2 1

Printed in the United States of America.

To the memory of my parents,
Poul Halling and Margit Halling

Contents

Acknowledgments

As this manuscript nears completion, I am mindful of the colleagues and friends who have so graciously contributed to this project over the past several years and whose encouragement has meant a great deal to me. Often, gratitude is described as a debt. However, for me, it is a blessing. It is a reminder of the care, energy, and interest of thoughtful people. I list them in no particular order.

Thank you to Jennifer Ferguson, my research assistant during the year I worked on the first part of the manuscript. She approached the task of reviewing what I had written, with intelligence, tact, and genuine interest. I am especially grateful to Mical Sikkema, who has patiently gone through a number of drafts of the chapters of this book and has helped me to become a clearer and more accessible writer. She has also offered much-needed perspective and encouragement on numerous occasions. Christie Lynk understood from the beginning what I was trying to do and has graciously and generously supported me along the way. Judy Dearborn Nill, fellow writer and friend, has been an invaluable resource in a number of ways, and especially as I developed my prospectus. She has also shown me that writing takes grit and persistence. Karen Lutz, our highly literate administrative assistant in the Psychology Department at Seattle University, has done a remarkable job of proofreading a major portion of this manuscript. My colleague Lane Gerber provided helpful input and suggestions on Chapter 2, especially on Heinz Kohut's theory. I am also grateful to Chris Mackie, who has patiently read and provided constructive criticism on drafts of each of the seven chapters. And Marie McNabb's comments on the first few chapters pushed me to think beyond what I had taken for granted. Also, thank you to Karen Litfin, who provided suggestions and encouragement during our quarterly walks around Greenlake in Seattle. I am fortunate to know Neil Elgee, the President of the Ernest Becker Foundation, and have benefited from his comments on my discussion of Becker in Chapter 6.

Thank you to Jeanette Valentine for reading and commenting on Chapter 5. Thank you, as well, to Mary Sacco for her help in constructing the index.

The Psychology Department at Seattle University is an unusually hospitable place. The faculty represents a broad range of approaches to psychology, and we work together as members of a genuine scholarly and collegial community. This is a rarity in academia, and I consider myself fortunate to be part of this department. There are two of my colleagues in particular whom I want to acknowledge. It was George Kunz's vision of a graduate program in existential-phenomenological psychology that originally brought me to Seattle University, and it is within the context of teaching in this graduate program that I developed most of the ideas expressed in this book. Since 1984 I have collaborated with my late colleague Jan O. Rowe on numerous research projects, including three studies on the psychology of forgiveness. I am deeply grateful to Jan for these years of dialogue and shared learning even as I mourn her passing.

Over the years I have benefited from conversations with my colleagues James Risser (Philosophy) and Michael Raschko (Theology) at Seattle University, Brayton Polka (History) at York University, and Thomas Trzyna (English) at Seattle Pacific University. These conversations have broadened my understanding of scholarship, and I am grateful to all of them.

Thank you as well to the members of PRAG House, the other community to which I belong. Over the years I have worked on this project, my fellow residents in this intentional community have shown their interest and support by inquiring gently, but not too often, as to how my book was coming along. My partner Kathryn Carson has also been a great source of support during the writing of this book. I am very appreciative of her understanding of how important this project has been for me.

I want to express my appreciation to the staff of Palgrave Macmillan with whom I have had the pleasure of working as this project has been brought to completion: to Amanda Moon, who was my first editor, and her successor, Brigitte Shull, both of whom have been gracious and supportive, to Katie Fahey, who has responded quickly and patiently to my many questions, and to the staff of Macmillan India, who have done much to improve the clarity of the manuscript through their attention to detail and nuance.

Finally, I am grateful to all of the people who have shared their stories with me. I only hope that I have at least begun to do justice to the richness and depth of their experience.

Introduction: On the Disappearance and Appearance of Persons

Intersubjectivity is experienced in a primordial way rather than known through proofs.

Maurice Natanson[1]

We all long for someone with whom we are able to share our peculiar burdens of being alive.

Salley Vickers[2]

To live is to live with other people. Typically, we assume that we understand the people who are important to us, and especially those with whom we are intimate. But how well do we really know these people, be they our friends or lovers, family members or colleagues, enemies or allies? No doubt we have some awareness of their significance to us, and yet it is easy to lose sight of them as persons with their own lives. Also, recognizing people's importance to us is by no means the same as understanding their point of view, knowing something of the world they inhabit, and appreciating what matters to them.

Then, suddenly, we are startled into taking a second look at someone whom we thought we knew intimately. Perhaps, in a time of crisis, we discover that a person we regarded as a close friend does not really understand us or that what we believed about this person no longer seems true. As George Steiner puts it, "We can be, in ways almost unendurable to reason, strangers to those whom we would know best, by whom we would be best known and unmasked."[3] Being surprised by those we thought we knew so well is a frequent theme in films and in literature. For example, Anita Shreve's best-selling novel *The Pilot's Wife* tells the story of a woman who begins to suspect that her husband, a pilot whose plane crashes into the Atlantic, had a secret life. Gradually, she is thrown into turmoil as she is confronted with information that does not at all fit with her picture of who he was.[4]

Such times or moments of surprise, of being thrown off balance, bring us back in a very personal way to fundamental questions about human relations: How is it possible for us to misconstrue someone with whom we have close contact, what does it mean to know another in an intimate way, and how does a deepening of a relationship come about? Is it even possible that two human beings can achieve a genuine measure of understanding? These are the sorts of questions that this book will address.

In exploring these questions, I want to engage both the heart and mind of the reader. This is not primarily a book about theories of human behavior or speculations about the human mind or psyche. Rather, I hope that the stories and reflections that I present will speak to the reader in a personal way, that they will resonate with what the reader has experienced, sensed, or intuited. In turn, this book owes its existence to a set of experiences, or awakenings, that I had when I was in my twenties. These awakenings led me to reconsider what it means to be a human being and to be genuinely open to another person. The sort of experiences to which I am referring is described in detail in Chapter 1, entitled, "Seeing a Significant Other as if for the First Time." Even now, more than thirty years later, I remember these events with a sense of gratitude and awe.

Yet one of the paradoxes of human existence is that although relationships can deepen and mature, they are often repetitive and restrictive. Our lives, we must acknowledge, are as much about frustration and disappointment (in ourselves and others) as they are about renewal and satisfaction. These realities will also be explored.

The stories and descriptions that I present are ones I have collected as part of my research as a psychologist. This book is based on evidence—the evidence of everyday life, the life with others in which all of us participate, even if we may be at a loss for words to describe the nuances and richness of it. In psychology and related disciplines there are many scholars who write about interpersonal relations (and, obviously, I am one of them). I believe these writings are of value first and foremost to the extent that they speak to our experience and thereby bring us to a deeper acknowledgment and appreciation of our own relationships, with all of their limitations as well as their possibilities for growth.

The tradition that values the evidence of everyday life and therefore believes that descriptions and stories are essential for understanding human existence is called *phenomenology*. It started as a movement within philosophy and, over time, it has influenced the various disciplines within the social sciences, including psychology. Later in this book (Chapter 5) I will discuss the phenomenological tradition and how I have drawn upon it. Its fundamental aim is to study human experience carefully and systematically and to express the insights arising out of such study in language that does justice to these experiences. Phenomenologists want to be faithful to the phenomena they study and thereby bring us back into contact with our experience, but in a new and fresh way—in an attitude of openness and wonder. This is the basic meaning of the often-repeated slogan "Back to the things themselves," formulated by the philosopher Edmund Husserl, the founder of phenomenology.[5]

Phenomenologists engage in three levels of analysis. First, they look at a particular experience, such as one person's story of being disillusioned. Once the person has described the experience in detail, the phenomenologist can reflect on what happened and, through this reflection, learn something about this particular event and the person who experienced it. For example, consider the case of a woman who has a one-dimensional and idealistic understanding of her doctor; it was this narrow view of the other that set her up for being disappointed. Moreover, her naiveté becomes more understandable as she describes growing up in a family where physicians are greatly admired.

A second level of analysis involves the search for themes that are common to a variety of experiences of being disillusioned (this is the focus of Chapter 2). Phenomenologists would compare this individual story of being disillusioned with other descriptions and start to elucidate what is essential to all of them. Such an analysis makes it possible for them to write about the nature of disillusionment in general. The third level is more philosophical and foundational. At this point, phenomenologists ask what it is about the nature of human beings and their relationships that make surprise and disillusionment possible and even inevitable. In a preliminary way, one could say that humans are continually engaged in constructing their future and their identity, and that looking up to selected others who are seen as powerful and remarkable helps to make this possible. That we are slow to become disillusioned reminds us of our extraordinary capacity for overlooking data that contradict our beliefs and expectations.

When philosophers, psychologists, and other social scientists reflect on experience, there is risk that they (and others) come to regard the theories they create as more "real" than everyday experiences, and that the joys and sorrows of flesh and blood human beings are left behind. In this regard, the Danish existentialist Søren Kierkegaard (1813–1855) once commented caustically on the abstractness of the philosophy of Georg Hegel, whose system of thought was dominant in nineteenth-century Western Europe: "It is like reading out of a cookbook to a man who is hungry."[6]

But phenomenology does not conceive of the relationship between lived experience and reflection as a one-way street, that is, as a movement from experience to abstraction or theory. The goal is not to construct a map of the world or of experience but to move back and forth between experience and reflection.[7] With this in mind, I turn to a story to bring us back to everyday human relations.

The English writer G. K. Chesterton's suspenseful and whimsical short story "The Blast of the Book" invites us to take a fresh look at our relationships to those people whom we take for granted in our lives.[8] Chesterton's story is fictional, but it has a truth at the core of it, a truth that is close to home.

In this story we are introduced to Professor Openshaw, a friend of Father Brown, Chesterton's legendary amateur detective, theologian at large, and philosopher of the everyday.[9] Openshaw is a scientist who specializes in the study of the occult and who takes great pride in his impartiality. He is as willing to dispute the faith of staunch spiritualists as he is to marshal evidence to challenge materialists who automatically dismiss the possibility of psychic phenomena. In his long career, he has unmasked many fraudulent mediums and has developed a reputation as a shrewd and careful observer of human behavior. Little does he suspect that his acuity as a scientific investigator is about to be put to the test under the most unusual circumstances.

In a brief talk that Openshaw has with Father Brown one morning, the issue of unexplained vanishings comes up. The two men agree that disappearances of people may be harder to explain than appearances of ghosts or fairies. As an afterthought to this conversation, the professor confides to Father Brown that a missionary is coming that very day to discuss some strange disappearances and he promises to tell the priest more at lunch. Professor Openshaw then hurries to his office, where he inquires of his clerk, Mr. Berridge, if his visitor, the Rev. Luke Pringle, has arrived. The clerk, who is busy adding up figures for the professor's various publications, replies in the negative and goes on with his work. Openshaw offhandedly gives Berridge additional instructions about the tasks he is working on and tells him to send Mr. Pringle straight into his private office as soon as he arrives.

While waiting for his visitor, Openshaw reflects on the letter that he received from Pringle several days before. It had impressed him favorably because it was an organized and straightforward typewritten account, not the kind of a letter a crank would send. Just at that moment, the author of the letter appears. Looking at him closely, the professor sees a man with a wild and untamed beard but humorous, sparkling eyes, the kind of eyes that inspire confidence. His manner of dress is shabby, but this does not seem odd, given that he is a missionary who has just arrived from Africa.

His visitor tells him an extraordinary story about an old book in a leather binding, first shown to him by a Captain Wales, one of the very few white people in the district in West Africa where he himself was stationed. The captain had gotten this book from a man he had met on a boat. This man had warned Wales that the book was extremely dangerous; anyone who opened it would be carried away by the devil or disappear. Wales had scoffed at this idea, and the two men quarreled. Prodded by Wales's expression of disbelief, the man with the book eventually did open it and look inside. He then immediately dropped it, walked to the side of the boat, and disappeared without a trace. Pringle tells Openshaw that he thought the story had elements that made it believable. What happened next only strengthened his belief in its veracity.

One day, when Pringle visited Wales in his tent, the captain started to talk about opening the book to get beyond his superstitious fear of it. At that moment, Pringle had his back to Wales as he was looking out at the jungle. Feeling suddenly uneasy, he cautioned Wales not to touch the book, but when he turned around, the book was lying open, the captain had disappeared, and there was a rupture in the back wall of the tent as if he had gone out that way. What could be more convincing, the missionary asks the professor, than what he had seen with his own eyes?

So now the Rev. Pringle has brought the book back to London to return it to its rightful owner, a mysterious Anglo-Indian doctor by the name of Hankey. Professor Openshaw is intrigued. While he would be inclined to dismiss Pringle's story as fantastic, the reverend does not seem to be the kind of person to tell tall tales. Where, Openshaw wants to know, is this mysterious book now? His visitor replies that he left the book in the front office with Openshaw's clerk because he feared that the professor might open it. And the clerk seemed like a solid type of person who would not look into other people's books. Openshaw is in complete agreement. "Your magic tomes are safe enough with him I assure you. His name's Berridge—but I often call him Babbage; because he's so exactly like a Calculating-Machine. No human being, if you can call him a human being, would be less likely to open other people's brown paper parcels."[10]

It is unlikely that the professor has ever made a pronouncement that was so immediately and emphatically contradicted. When the two men enter the outer office, they find that the book has been unwrapped, Berridge is nowhere to be seen, and the large picture window in front of his desk has been shattered as if the clerk somehow jumped right through it.

Completely taken aback, Openshaw tells Pringle that this latest disappearance has eradicated from his mind any doubts about the veracity of the missionary's story. When Pringle suggests that they make some inquiries about the clerk's whereabouts, possibly calling his home or notifying the police, the professor admits that he does not know where Berridge lives, or whether he has a phone, and that he has never really noticed what he looks like, except that he is rather ordinary and wears glasses. The Rev. Pringle then suggests that he take the book to its rightful owner, Dr. Hankey, for his opinion. After some hesitation, Professor Openshaw reluctantly agrees to this course of action. When Pringle returns a short time later, without the book, he tells the baffled and dazed professor that he has left it with Dr. Hankey, who wanted to study it for an hour. The doctor had urged that both of them come to see him subsequently. The professor is dubious about this doctor who is completely unknown to him, but agrees once more to follow Pringle's lead. Before leaving, he calls Father Brown and arranges that they meet for dinner rather than for lunch.

At this juncture, the reader can readily anticipate what these two men find when they arrive at the doctor's house: the book has been opened, and there are signs that the doctor has run out the back door and vanished into thin air. Apparently Dr. Hankey has also made the wrong decision. Not knowing what else to do, the men proceed to the restaurant where Father Brown is waiting for Openshaw. As soon as they get there, Pringle quickly puts the leather bound book on a small table. After drinks, the missionary asks if he may be allowed to use Openshaw's office to sit and think through the whole mysterious business. The professor agrees, and Pringle departs with the book.

When he turns back to Father Brown, Openshaw finds the priest talking to the waiter about his young baby who had recently been sick. In surprise, Openshaw asks how Father Brown has come to know the waiter, and the priest casually replies that since he comes to the restaurant several times a month, he occasionally converses with him. The professor, who eats at the restaurant five times a week, suddenly realizes it has never occurred to him to ask the waiter anything about himself. This train of thought is disrupted when he is called to the phone. It is Pringle, who in an emotional tone tells him that he has decided to open the book and that Openshaw can't talk him out of it. This is followed by an eerie sound, and then the phone is silent.

When the professor returns to his dining companion, he tells Father Brown the whole story of the five men who vanished, adding that he is most puzzled by the disappearance of Berridge because he was such a quiet fellow. Father Brown comments that Berridge is indeed a very conscientious man who keeps his work and his personal life so separate that hardly anybody knows that he is a humorist at home. The professor is astonished to learn of the priest's familiarity with his clerk. Father Brown matter-of-factly tells Openshaw that he has conversed with Berridge on those occasions when he was waiting for Openshaw at his office. From these conversations he also has learned that Berridge likes to collect odd items of little value.

This adds to the professor's astonishment, but he exclaims that it does nothing to explain the man's disappearance or the disappearance of the four others. This is when Father Brown provides his explanation:

> "Berridge did not disappear," said Father Brown. "On the contrary."
> "What the devil do you mean by 'on the contrary'?"
> "I mean," said Father Brown, "that he never disappeared. He appeared."
> Openshaw stared across at his friend, but the eyes had already altered in his head, as they did when they concentrated on a new presentation of a problem. The priest went on:
> "He appeared in your study, disguised in a bushy red beard and buttoned up in a clumsy cape, and announced himself as the Rev. Luke Pringle. And you never noticed your own clerk enough to know him again, when he was in so rough-and-ready a disguise."[11]

The professor is not yet convinced. Father Brown points out that Openshaw paid so little attention to Berridge, thinking of him as little more than a calculating machine, that he could not describe him to the police. Moreover, he knew so little about his clerk that he would never have guessed he was capable of staging an elaborate prank. Since Openshaw had never been to Berridge's house and had no idea where he lived, it was easy enough for the clerk to take one of the odd items he had collected, a name plate for a certain doctor, and put it on his front door, and lo and behold, "Dr. Hankey's residence" came into existence.

And since Father Brown is not in the least bit superstitious, he took a quick look at the dreaded book while "the Rev. Pringle" and Openshaw were having drinks. On the outside it carried a warning, in three languages, of dire consequences for anyone who opened it. On the inside, there was nothing but blank pages!

Why would Berridge stage such an elaborate prank? the professor wonders.

> "Why, because you had never looked at him in your life," said Father Brown. . . . "You called him the Calculating-machine, because that was all you ever used him for. . . . Can't you understand his itching to prove that you couldn't spot your own clerk?"[12]

Father Brown admonishes his friend not just to look at potential liars and fakes, but also to look at ordinary, honest people like the waiter or his clerk. And Openshaw is big enough to take in what the priest tells him and laugh at his own foolishness.

As readers we are also inclined to laugh at Professor Openshaw, forgetting that we were, very likely, as spellbound by the melodramatic story that "Pringle" recounted as was Openshaw. Therefore, we might think of ourselves as rather like Father Brown who, unlike the professor, takes time to get to know people on a personal basis. That is, we can think about this story in terms of the first level of analysis, discussed earlier, considering it just in terms of the particular incident and specific characters. But, if we think about it more, we have to acknowledge that all of us overlook other people in our lives, those who wait on us in restaurants, work for us, or serve us in some other capacity. And it is not just that we overlook those people with whom we have a functional relationship, such as our employers or employees. How much attention do we give to those people who are most significant in our lives and with whom we have a personal relationship? How open are we to them and they to us? How curious or interested do we allow ourselves to be? Here we are at the second level of analysis, thinking of this phenomenon of overlooking others as it occurs for all of us and in various situations.

Our overlooking of others is not necessarily or even typically "conscious" or deliberate, of course. The professor did not realize he knew so little about his clerk. He thought he knew who this man was: the quiet clerk who did his work and would not dream of opening a brown package, let alone set in motion an elaborate production that would thoroughly unsettle his employer. And how little did Openshaw realize the extent of Berridge's understanding of him; for example, Berridge knew quite specifically what Openshaw thought of his clerk. Yet what the professor believed was not entirely erroneous. His clerk did work diligently, was relatively quiet, could be described as ordinary looking, and did wear spectacles. But what Openshaw did not see—the sparkling, humorous eyes, his soul, as Chesterton puts it—was really at the heart of the matter.

In this story, it is the action of the person who is overlooked that brings about a change in perception and thus a transformation in the relationship. And surely it is no exaggeration to say that the relationship has been transformed. However much Berridge remains the clerk and Openshaw the professor and employer, we can assume that Openshaw will not be able to look at (and overlook) Berridge in the same way that he had previously. Although he has not yet really come to know Berridge by the end of the story, he has certainly glimpsed enough of the creativity of his clerk to recognize the obvious inadequacy of his previous image of him. What Berridge has accomplished through staging the fantastic episode of the "dangerous" book is to dramatically change the circumstances of their being together. It is not just that the clerk became somebody else—the Rev. Luke Pringle. Instead, for a short period, the context of their being together altered fundamentally. Berridge, disguised as the Rev. Pringle, brought to the professor a fascinating story pertaining to the latter's area of interest and was in control of the drama that ensued. It was within these new circumstances that the impetus for a transformation emerged. Had Father Brown suggested to Openshaw that he pay attention to honest, ordinary people prior to this whole dramatic episode, his words would likely have been without effect. Openshaw's conviction that he was an expert in observing people and in uncovering what lay beneath the surface would surely have warded off even the most cursory self-examination. There would have been no shared experience to give meaning to such a suggestion, no disruption of the taken-for-granted assumptions resulting in humility and new learning. Here we are getting closer to, if not quite into, the third level of analysis where the question becomes the nature of human beings, their consciousness, and relationships.

Chesterton's story illustrates how much one's perception of others depends on the context in which one sees them. This point was brought home to me some years ago when I at first failed to recognize someone I knew quite well. I entered a movie theater, and the usher who took my ticket greeted me by name. My concern, up to that point, had simply been to orient myself—the theater was relatively dark and I was not wearing my glasses. Little did I expect that I would meet anyone there, least of all an usher, who would know me. All of a sudden I was in the embarrassing position of being addressed by someone who obviously knew who I was, and whom I did not recognize. When this woman repeated her name (initially I had not been able to hear what she said any more than I could see who she was), her face finally came into "focus" for me. There *she* appeared, this person herself, right in front of me. She was a former student for whom I had a great deal of respect. Once her name registered, so also did her face, and I immediately remembered the last class she had taken from me and something of her interests. Had we met on campus, I would have recognized her immediately.

The first chapter in this book addresses the vital role that the particular contexts in which we interact with people play in sustaining our habitual perceptions of them and, conversely, in creating opportunities for coming to experience them in deeper and more intimate ways. This chapter is pivotal in that it points to the possibility of depth and understanding in human relationship; hence the reference to the experience of "intimacy" in the title of this book. Intimacy conjures up images of close relationships with friends or family members, for example. However, as we know, these relationships are not necessarily intimate in the etymological sense of having to do with the inmost part of oneself or the other. (The word intimacy is derived from the Latin, "intimus," meaning "inmost.")[13] I am referring to experiences of close contact, whatever the "official" nature of the relationship. The question is not whether the stories or descriptions use the word "intimacy" to describe what happened, but whether the story has the quality of intimacy. The following story provides a strong example of what I mean by intimacy.

Denise,[14] a young woman, wrote of how she came to see her husband Gary in a new way after four months of marriage.[15] He was the director of a choir consisting of employees from the company where he worked as an accountant. During the first Christmas after their marriage, this choir was invited to sing at a special dinner dance, and Denise decided to attend. This was one of the first times during their marriage that she saw Gary outside their home—they rarely went out, because of limited funds. Denise described the change that occurred in her view of Gary as follows:

> Tonight . . . I not only saw Gary in a new setting, but I saw him in an entirely new context. Not only was he my husband—he was the [choir] director, he was an accountant, he knew and he had respect from all these people with whom he worked everyday and I saw Gary in an entirely new way. Maybe I saw him as his own person—as an individual and how he relates to others instead of as only how he relates to me. I watched him as he spoke with others, how he acted around and with these people I didn't know, and as I watched him direct a great performance, I realized that this was a part of Gary's life which I was not included in. I don't mean to say that I felt excluded
> —only that I was observing something entirely different from what we shared. . . .
>
> I can remember carrying with me throughout the remainder of the evening the strangest feeling, that in a sense, I had met someone new. It may sound corny or trite to say this, but in some odd way it was true . . . I suppose that on that evening I came to realize that you can never really discover all there is to know about another human being who holds real significance for you, even though you live with that person and interact with them every day.

The change in context provided Denise with the possibility of coming to a new appreciation of her husband. Even though she hints at some discomfort in this situation as she realized that there was a part of Gary's life that did not include her, she allowed this profound realization to sink in. She acknowledged that one never comes to know another person in a definitive way and that one's own life and that of another can never fully coincide.

Although the focus of this book will primarily be on "significant" relationships—on our assumptions about and perceptions of those close to us and on how these are transformed—the whole question of how we recognize or fail to recognize something of the personhood of fellow human beings has implications far beyond the realm of the interpersonal and the psychological. A powerful example of such an implication is given by Alfred Speer, Hitler's minister of armaments and munitions during World War II.

In one of a series of interviews with the journalist Gitta Sereny, Speer describes how his attitude toward the whole war effort changed in the spring of 1944 as it became obvious that the war was going to be lost. He was on the verge of resigning from his position because he thought that a number of Hitler's decisions, such as initiating a project to build six underground industrial sites, were completely unrealistic. Walter Rohland, one of the industrialists with whom Speer had worked closely and become friendly, urgently appealed to him not to resign. Who else, Rohland asked, could deter a desperate Hitler from pursuing a scorched-earth policy as the Allies moved into Germany? The specter of Hitler ordering the self-destruction of Germany took Speer aback. He told Sereny,

> I don't know how to explain it. But for the first time, I think, I stopped thinking of myself and thought only of our country—of the people. You know, all those terrible months in 1943, when on my many trips I saw so much destruction, can you believe that I never thought of people then? Of what it was doing to them? I only thought of my damned factories. It was as if imagination had died in me . . . suddenly for the first time in years, I had a sudden vision of physical destruction—not of buildings, but of people.[16]

While Rohland and Speer were having this conversation, Speer's children played around him while overhead was heard the noise of Allied planes flying on bombing raids over Germany. As Sereny suggests, the immediacy of this threat to members of his own family might have helped Speer to at least begin to imagine the horrible reality of further destruction of human life.

Unfortunately, we do not have to turn to Nazi Germany—or wartime, for that matter—to find examples of how readily one loses sight of other persons as persons, a lapse that has terrible consequences. A recent example comes from the sphere of globalization and economic assistance, so called.

Joseph Stiglitz, winner of the 2001 Nobel Prize in Economics, has written a scathing indictment of how the International Monetary Fund (IMF) has treated developing nations, showing no apparent regard for the effect of their stringent "free market" loan requirements on ordinary people in these countries. The following is his pithy analysis of what enables the staff at the IMF to carry out such policies. He notes that the IMF (in contrast to the World Bank) does not have staff living in the countries where they do business. Rather, IMF representatives fly and in and stay at luxury hotels and then fly back out after a brief visit. The problem with this way of dealing with economic realities, he argues, is that "one should not see unemployment just as a statistic, an economic 'body count,' the unintended casualties in the fight against inflation or to ensure that Western Banks get repaid . . . Modern high-tech warfare is designed to remove physical contact: dropping bombs from 50,000 feet ensures that one does not 'feel' what one does. Modern economic management is similar: from one's luxury hotel, one can callously impose policies about which one would think twice if one knew the people whose lives one was destroying."[17] Of course, many times, the person whose life one is destroying may be quite close—just a few offices down the hall from one's own.

Chapter 1 of this book focuses on a phenomenon I call "Seeing a Significant Other as if for the First Time." Based on numerous descriptions of transformations relationships, such as the one provided by Denise (given earlier), this chapter considers the basic dimensions and consequences of such an experience and provides an interpretation of what makes it possible. It takes us into "positive" experiences of change, where the other is seen as "more than" who one thought he or she was. Such experiences set the stage for the development or deepening of closeness. In contrast, Chapter 2 deals with disillusionment, in which the other who is so important to us is found to be "less than" who we thought he or she was. This is a profoundly unsettling experience, one that often leaves one at a loss as to how to proceed with one's life. Disillusionment not only shatters one's image of the other but also raises troubling questions about oneself and one's own history. This chapter deals, in a sense, with the loss of intimacy. How one responds to and attempts to come to terms with disillusionment is a key issue that is explored in detail.

Chapter 3 looks at one possible direction to coming to terms with disillusionment, namely, a movement in the direction of forgiveness. At its core, the experience of forgiving another is an experience of coming to acknowledge the basic humanity of the person who has injured us; implicitly, it also involves a deepening acceptance and recognition of our own humanity and fallibility. In other words, forgiving another implies what is commonly called "self-forgiveness."

Understanding and appreciating the point of view of the other is a core aspect both of forgiving and of coming to see the other more fully in his or her distinct personhood. Chapter 4, "Experiencing the Humanity of the Disturbed Person," addresses the question of whether it is possible to come to a genuine personal understanding of disturbed persons. If so, how is this possible, and what are obstacles to such an understanding? We will see that the answers to these questions take us back, from a different angle, to the experiences discussed in Chapter 1.

While the first four chapters focus on human experience and stories, on what it is like to be disillusioned, to forgive, to see another as if for the first time, and to see the basic humanity of the person who is disturbed, the second half of the book raises some broader theoretical issues and questions. Chapter 5 provides an overview of the phenomenological tradition that provides the method and guiding philosophy for this study of human experience. As we have seen, this tradition is concerned, above all, with doing justice to the ambiguities and nuances of human experience.

Chapter 6 examines the deeper implications of the kinds of experiences treated in chapters one through four under the heading of "Interpersonal Relations and Transcendence" in everyday life. It clarifies what is meant by transcendence and argues that it is this fundamental but typically overlooked or even denied aspect of our existence that is at the very core of experiences of being surprised "by the other." The term transcendence is often used with otherworldly connotations, as if to transcend is to leave the realm of the actual, or as if transcendence is a quality that belongs to the divine. In contrast, I argue that our openness to possibility means that transcendence is at the very heart of our humanity: *our experiences of intimacy* (as well as forgiveness and other related phenomena) *testify to the reality of transcendence in our lives.*

In Chapter 7, entitled "Psychology, Transcendence, and Everyday Life," I review some of the contemporary psychiatric and psychological literature on interpersonal relations, considering its positive contributions as well as its limitations in light of the findings and reflections that form the body of this work. Finally, in light of psychology's limitations as a guide, I address the question of where else and how we should look to deepen our appreciation of the depth of human relationships.

I have written this book hoping to invoke the reader's own experience of the interpersonal, and, for this reason, I plunge directly into a discussion of descriptions of events within relationships. Those readers who would first prefer to know something of the basic approach taken in this book are encouraged to read Chapter 5 before starting Chapter 1. Otherwise, the chapters are most readily understood when read in sequence.

CHAPTER 1

Seeing a Significant Other "As if for the First Time"

And do you know, Maia, he actually looked at me, really looked, and it seemed to me he was then seeing me for the first time.

C. S. Lewis[1]

But sometimes stories are all we have. Sometimes they are all we need.

Jeffrey Smith[2]

Introduction

The experience of genuinely being seen by another person is one that we deeply long for. What could be more wonderful than being recognized for who one really is by a lover, parent, or friend? And yet we are also keenly aware that becoming visible to another is risky. Think of a time when you were feeling vulnerable, ashamed, or even terrified at the prospect of being seen—even by someone who deeply cares about you, let alone someone who would judge you.

We go to great lengths to protect ourselves by hiding from others (and from ourselves) many of our deepest thoughts, feelings, and inclinations. Yet, while we hide, we also keep an eye out for someone who might care about us and value our point of view, or, to use a psychological term, someone who might empathize with us. As the psychoanalyst Phil Mollon has written, in presence of empathy we may feel safe enough to expose and express who we are,[3] but perhaps still with a sense of trepidation.

The reverse side of this experience is that of becoming fully present to and seeing another person in his or her depth and complexity. This too, as I suggested in the Introduction, is an extraordinary moment in one's life and in one's relationship with the other. When I was in my early twenties I had several such "interpersonal epiphanies" that touched me in a profound way.

They changed my life and provided the impetus for most of the writing and research I have done as an academic, especially the writing of this book. More importantly, they gave me a new way of looking at the world.

It is both strange and regrettable that such experiences are barely addressed in the psychological literature. In contrast, researchers in the field of communication have been studying "turning points" in relationships for some time. For example, L. A. Baxter and C. Bullis have interviewed partners in romantic relationships to determine what they regard as turning points in the movement toward greater intimacy.[4] These researchers discovered, as have I, that the people they interviewed could readily recall and describe such events in detail.

To gain a deeper understanding of such moments of epiphany (or recognition), I have solicited first-hand descriptions from a variety of people, asking them to either write a description or tell me their story in an interview.[5] Of the sixty-five descriptions I have collected, seven were obtained through lengthy interviews.[6] I asked each person a question that is both simple and direct:

> *Describe as specifically as possible a time when you came to see someone of real significance in your life more as a real person in his or her own right.*

I used this question because it worked: virtually every person to whom I put this question was able to respond with a story about coming to see someone of importance in his or her life "as if for the first time," along the lines of Denise's description in the Introduction.[7] There I suggested that all of us, in some measure, take for granted or, in some other way, overlook other people, including those who are most important in our lives—family members, lovers, friends, coworkers, or even adversaries. Based on the assumption that we really know who the other person is, we are often quite inattentive, even if not as dramatically as was Professor Openshaw with regard to his clerk, Mr. Berridge. Nonetheless, any experience of coming to see the other "as if for the first time" is likely to be a milestone in one's relationship with that person and thus a memorable occasion.

One undergraduate student wrote an especially moving account of such a turning point. The incident she described, which took place when she was thirteen, brought about a transformation in her relationship with her mother and changed both of them.

Mary was the second of three children in a large family. She had a very close and yet sometimes difficult relationship with her mother. Together with the other two young children, Mary spent much of her time with her mother, especially before she started school. These were largely enjoyable times. The four of them would take walks, work on arts and crafts, and sing songs together. The main problem for Mary was her mother's occasional temper. As Mary put it, "She could flare up into a blinding rage over matters which seemed trivial to me. And if I couldn't understand what she had been upset about, she would get angrier." Mary even used the word "monster" to capture her experience of her mother when angry. However, unlike her siblings, Mary did not run away from her mother when she became upset.

As the years went by, it became increasingly important for Mary to defend herself against the accusations her mother directed at her during moments of anger. When she thought back on these painful incidents as well as her overall relationship with her mother, she realized that she had considered herself somehow responsible for her mother's behavior and thus felt obligated to do something to improve the situation. Moreover, she was concerned about the effect these episodes of anger were having on the family.

For years mother and daughter lived almost as the best of friends. At the same time, Mary was often on the edge of wariness as one small incident could upset her mother and set her off "on a long rant of I've-been-picking-up-the-house-for-over-twenty-years-and-I-don't-know-what-I'm doing-here." Finally, there came a break in the pattern:

> One day, I was thirteen and in the middle of the "nobody-appreciates-me-for-twenty-years-of-giving-and-giving" broken record when it suddenly dawned on me just how long twenty years is. It was longer than I had been alive. She had been fighting the house, doing work she didn't care for with no one but little kids for company for over twenty years. I looked at her amazed, and the monster melted into a frustrated middle-aged woman who began to cry. I understood. I started crying too, and she cried harder because she realized that someone understood her problem. I finally saw the real person, and I haven't seen the monster since. Now she's teaching other women to break out of frustrating lifetime situations. She doesn't feel as trapped by the house or her own situation and I finally understand her when she does get upset. We still fight, but we listen to each other through our anger. It is nothing like our helpless, frustrating, stagnant fights of old.

At first sight it would seem that this remarkable shift in Mary's fundamental attitude toward her mother occurred under the most ordinary circumstances— it is hard to imagine anything more pedestrian than the never-ending arguments that so often exist as a thread of continuity in many of our close relationships, especially with family members, spouses, or lovers. Few things in life seem as inexorable as these recurrent patterns of dispute and tension from which we seem unable to find any means of escape. "The next time this sort of thing comes up, I will respond differently and so something different will happen between us," we say to ourselves. But, of course, the next time we are drawn in exactly as before, our good intentions reduced to empty and impotent gestures.

Previously, I have referred to the necessity of understanding change in terms of the context within which it occurs. Mary's description is not exhaustive enough to tell us what was different about this incident, or whether something out of the ordinary had occurred in the previous hours, days, or weeks. But the concept of *context* reminds us that our actions are not determined by outside factors independent of our relationship to them. Context is not the same as the situation as it can be described by any competent observer. Within the tradition of phenomenology, context refers to the situation as it is given to the perceiving person in terms of the explicit and implicit meaning it has for him or her.[8] Context is a matter of how one is in relationship to a situation, how one understands it. And this meaning can change as a function of one's approach to the situation. For example, if I come to the realization that a colleague's brusque way of asking questions is a masked expression of uncertainty, then I no longer take his manner of asking questions personally, and the context of our being together changes in a fundamental way. For this realization to have occurred, something in me must have shifted such that I listened to this man in a different way from before, now noticing features of his expression and speaking that on other occasions had eluded me. This example also shows that the idea of context includes subtle and yet significant aspects of a person's experience of a situation.

Along this line, a young woman gave a wonderfully poetic description of how she came to see her boyfriend's brother in a new way, under circumstances that at first seemed quite unremarkable. Rachel started her story by stating that this person, Wayne, sounded a lot like his brother, Michael, and even resembled him physically. Both Michael and Rachel tended to worry about Wayne, thinking of him as a young kid who was apt to get into trouble because he was carried away by his enthusiasm. One evening Rachel went over to visit Wayne because she had spoken to him on the phone and he sounded discouraged. She had arrived at his apartment and, as she was looking through his refrigerator, hoping to find something for the two of them to eat, the phone rang. "I tried not to listen to what was happening on the phone," she said. "I turned to close the door and ask Wayne where a skillet was and stopped. The light from the kitchen fell on him for a second and in that second I saw Wayne not as Michael's brother but as Wayne." Later in her account, she elaborated, "The light falling on his face shadowed his eyes and brought out his cheekbones and his nose and sculptured his face. For that second I seemed to see what he was and what he could become. A strong man given to excess which could be controlled when he found what or who he was looking for." Wayne was speaking to a woman he was dating, and Rachel noticed both his puzzlement at what he was hearing as well as his gentleness in responding to his caller.

Rachel was well aware that it was not the light by itself that brought about this shift in how she looked at Wayne. But, we might say, the light falling on his face in such a striking way provided the means for her to take in what she was now ready to acknowledge—more of the fullness of his humanity, his existence as someone who had his own hopes and desires. She had the opportunity to observe him, not in relation to herself but as he spoke to someone else, and, at that moment, she saw him differently and in a more three-dimensional way. Although at one level everything is the same after such a shift in perception and understanding, at another level everything is different because the meaning of the other's behavior for oneself has changed in a very real way. Again, context is a term that emphasizes that the situation and person must be understood in their fundamental interdependence rather than as separate units. That is, the notion of context refers to a situation in terms of the personal meaning it has for the individual responding to it.

This change in personal meaning is evident in Mary's account. Just as she was in the middle of listening to "the broken record" of her mother's complaints, she suddenly heard these complaints differently than she ever had before. She took in what her mother said in a new way, imagining how long twenty years is, recognizing that it was longer than she had lived. That is, Mary related what her mother said to her own life and to her sense of her own history, and, by means of thinking this through, she came to grasp something of her mother's point of view and of the larger context of her mother's frustrations. Rather than thinking, "Twenty years—I can't even imagine that," she intuited what this statement meant in terms of the other's life experience by relating it to the length of her own life. She overcame or transcended the egocentricity of adolescence, not by disregarding her own point of view but by acknowledging it and thereby going beyond it.

Such a shift is not, one would assume, possible until preadolescence. The American psychiatrist Harry Stack Sullivan has suggested that not until eight or nine is a child capable of developing a genuine concern about what matters to another person, and that this concern first develops in relationship to a friend or "chum."[9] This is not to deny that the young child can discern that the other is troubled or in pain and respond with a gesture of kindness. But to grasp the point of view of the other in such a way as to understand the meaning of the pain as well as something of the other's self-understanding (as distinctive from one's understanding of him or her) is quite another matter.

While acknowledging the "cognitive development" that this openness to the other presupposes, we should also note that this experience was a deeply personal one for Mary. She grasped the reality of her mother's situation not as an "impartial observer" but as a deeply engaged daughter and fellow human being who cried with her mother. Yet, a number of thinkers in the social sciences, Sullivan among them, have assumed that a mature or more objective awareness of another person involves "scientific detachment."[10] None of the descriptions I have collected support this assumption. Rather, it appears that the deepest awareness of another is simultaneously personal and objective.

This interconnection of the personal and the objective is also evident in a related set of descriptions that I have gathered. For the last six years I have asked graduate students in one of my classes to write a description about coming to understand another person more objectively and realistically. These students have described changes in relationships with clients, colleagues, spouses and other family members, neighbors, and friends— people they knew well and people they had met recently. In each case, the student described a relationship where he or she had a commitment to, or at least an ongoing interest in, the other person. The students wrote about coming to an understanding of the other that, as in Mary's story, was much more three-dimensional than was their previous view. For them seeing the other more realistically meant that they came to understand the other person's conduct in terms of his or her life rather than just taking it personally or interpreting it more intellectually as an expression of a personality "trait." For example, one student wrote about her relationship with a colleague who was prying and confrontational. As she got to know her colleague better and learned something about her personal life, she recognized that this controlling behavior was an expression of how desperately out of control her colleague felt. Her behavior was not directed at the student as such nor was it a fixed behavior pattern caused by some internal trait. Rather, this was how she attempted to cope with overwhelming problems.

This student's simultaneously dispassionate and empathic view of her colleague came out of a series of prolonged and intense interactions with her, and not from just sitting back and thinking about her in an attitude of "scientific" curiosity. Such a process of ongoing struggle and engagement is typical of the over one hundred descriptions I have collected over the years from these graduate students.

In the moment that Mary grasped, from her perspective, her mother's point of view, everything changed. She was no longer primarily the recipient of her mother's anger or the person who was somehow responsible for making the situation better. The situation was not about her and what she had done wrong—it was about her mother and her frustrations. Furthermore, when her mother appeared, just as the "monster disappeared," she appeared not just as her mother but as a fellow human being. In a sense, Mary moved alongside her mother, seeing something of the world in which her mother lived. As the psychologist Louis Sass has argued in his discussion of the interpersonal, understanding another is not a matter of "getting inside someone's head" or achieving a "communion of souls," but of looking at what the other looks at, or, of arriving at a common meaning.[11] For Mary to understand her mother, she had to grasp something of what it was like for her mother to be home alone, year after year, with so many small children.

The historical background Mary provided helps us to more deeply understand this shift toward "recognition of the other" as well as its consequences for her ongoing relationship with her mother. She wrote that she had been very close to her mother and had spent a great deal of time with her, and that for the most part these times had been enjoyable. During her mother's outbursts, she had been the daughter who had stayed with her and had listened to her, however defensively. Perhaps her mother had hoped that some day Mary would in fact understand what her life had been like. In any case, a foundation had been created for the emergence of such an understanding through their mutual involvement and the time they had spent together in activities and in conversation. And then, through this epiphany, Mary's mother finally felt understood as she and her daughter cried together. Her mother was no longer alone with her burden and her frustration; their relationship had entered a new phase, both in terms of Mary's perception of her mother ("I haven't seen the monster since") and in terms of the manner in which they argued ("we listen to each other through our anger"). In addition, her mother's view of her own situation also changed ("she doesn't feel as trapped by the house or her own situation").

Feeling really understood by another person is not something that we are apt to take for granted. Only with a few people do we have the sense that we are really being listened to, that the other person is genuinely attentive to what we say and what we mean. We live in a world where it seems that everyone is rushing from one appointment to another and where we rarely give full and sustained attention to a fellow human being. On the other hand, when we do speak to someone about an issue that is important to us, it is disappointing when we see that the other person does not really grasp what we are saying and does not even seem to recognize that this is the case.

Why is it so important to be listened to? The answer may seem obvious: we feel valued when someone pays attention to us and what we have to say. But there is more to it than that. A psychological study conducted by Adrian van Kaam helps us to more deeply appreciate what is involved in feeling understood.[12] Based on descriptions collected from hundreds of research participants, van Kaam has articulated the basic character of this experience. My intention here is not to summarize all aspects of his analysis but to focus on several dimensions that are especially relevant to what happened between Mary and her mother.[13] Being really listened to means that one feels understood at an emotional as well as at a cognitive level. It means that one feels understood as an individual by a fellow human being who is actively attending to what one is saying. The genuine interest and care shown by the listener conveys acceptance of oneself and overcomes one's sense of being alone. Mary's mother was rarely by herself and yet, with just her children around her, she felt very much alone. Within this context of safety, there is the possibility of an experience of a deep personal relationship with the listener.

Van Kaam's elaboration of this last point is especially significant. He first states that there is an experience of safe, experiential communion with the one who is listening, and then he adds "and with that which the subject perceives this person to represent."[14] What does this mean? Van Kaam is suggesting that in connecting with this other person and feeling accepted and understood by him or her, one also experiences a communion with the larger world that the person represents. What the other person represents varies, of course, depending on who this person is and how one views him or her. Imagine, for example, a high school student who regards himself as an outcast and who then comes to feel understood by another student who is a member of "the in-group," or a woman who, as she grows older, feels excluded or left behind and who finds, much to her surprise, that she is understood by a younger woman.

The key point is that this experience of being understood by another person often has a meaning that extends beyond this particular event. This experience takes us beyond the realization that "at least one person understands me" to an affirmation of oneself as a member of the human community. The presence of this sense of communion with something beyond oneself and one's immediate listener is evident in the case of Mary's mother. Feeling understood by her daughter gave her a profound sense of relief and, eventually, led to a change in the overall direction of her life as she started to work with other women in similar circumstances.

Mary's poignant description and my preliminary reflections on it are intended to make clearer what I mean by "seeing a significant other as if for the first time." Such an experience is pivotal in any relationship and, in turn, must be understood within the context of the whole relationship. That is, these shifts bring with them something genuinely new while they are also rooted in the history of the particular relationship.

The Central Character of the Experience

I have discussed, in some detail, one particular story of interpersonal transformation. Now I want to consider what these experiences have in common, what features are particular to this "seeing of the other." (This is the second level of analysis that I discussed in the Introduction.) Based upon careful reflection on all of the accounts I have collected, attending both to what is expressed implicitly and to what is stated explicitly, I have concluded that the fundamental character or structure of the experience can be described in terms of five interrelated constituents or themes: (1) surprise and wonder, (2) participation in the perspective of the other, (3) recognition of separateness, (4) awakening of the self, and (5) a horizon of hopefulness. These themes are distinct and yet inseparable. That is, they can be discussed one by one while each one is inextricably connected with all the rest.

All of our experiences, and most evidently those that involve fundamental changes in our lives, are inexhaustible in their richness. Nevertheless, we can illuminate central features of phenomena by approaching them from different angles, holding up to the light one aspect at a time. Simultaneously, we should acknowledge that there is always more to be said, and that truth and a deeper understanding are served by a variety of perspectives and approaches. For each theme I use specific stories as illustrations. This does not mean that the theme fits only the stories mentioned, but rather that the particular story or stories selected exemplify the theme in question especially clearly and vividly.

1. Surprise and Wonder

The words "surprise" and "wonder" emphasize that the experience was not one that we had been deliberately working toward, that we were deeply touched by whatever or whoever we were surprised by, and that our assumptions or expectations were brought into question in the face of some new and unanticipated reality.

Carmen, in writing about an incident involving herself and her four-and-a-half-year-old son, has captured this sense of surprise and wonder vividly:

One day when I was at the kitchen table studying, Davey came to me with his arms folded over his chest, and a stern look on his face, and said, "I'm angry at you mommy." It was a blow to me that my son was angry at me. How could he be angry with me? How dare him. These questions raced through my mind in a matter of seconds. I stopped everything and gave Davey my full attention and responded. "Why are you angry at me, Davey?"

"Cause you made me kiss you good-bye today in front of all my friends [this was at school]. I'm too big for that, mommy," Davey said.

"What do you suggest I do, Davey?" It took him a while to answer me, but he ended up suggesting that we shake hands.

"WOW!!!" I said to myself. Shake hands! This is my baby talking to me. Before I could get a good grip on my emotions, Davey had said, "Don't be sad, mommy. Only babies do that, and I am not a baby."

. . . I realized then, and from that point on, Davey had a right to react in his own way, to objecting to the way I do things. He had a right to express himself whether it would make me happy or sad, because he was an individual, a person in his own right, that was learning to think a little on his own, and feeling concerning certain subjects.

. . . Now, that I have this "light," I try very hard to get to know Davey, and the more information I find out about him, the more my love develops for him.

This metaphor of "seeing the light" was used by a number of the people—including Rachel (mentioned earlier)—who told me their stories. It implies that they looked back and realized that they had been in the dark, unaware of the depth of the other person. Moreover, by definition, light is both that which we cannot help seeing (unless we close our eyes) and that which makes seeing possible. In this account, as in Mary's, surprise, discovery, "seeing the light" fully engaged the person: to be surprised is literally to be taken by surprise, to be present to or to be overcome by what appears. Thus, we can see that surprise is intimately connected with the theme of awakening of the self. That is, habitual patterns of perceiving and responding give way to a freshness of experience and to a deeply personal and spontaneous reaching out to the other.

This reaching out does not come about as we make a deliberate effort to bring about a certain kind of transformation. Carmen was sitting at a table studying, having no anticipation of what her son would say to her. As she listened to her son, she responded to him from a very deep level within herself, a level much deeper than that of her conscious will. Davey had asked something of her that was much more demanding than helping him with his homework or some other task. He was asking for her complete attention; he wanted her to realize that he had changed and that therefore their relationship must also change.

The philosopher Martin Buber has written compellingly and insightfully about the way we are challenged to grow through our relationships. One of his most basic distinctions is between an "I-it" and an "I-Thou" relationship. The former describes those interactions with others that are primarily functional in nature, that is to say, the relationship is a pragmatic one where the other is seen as a means to an end. Our business relations are within this mode, but the I-it relationship also is part of our dealings with friends and family. For example, parents invariably experience their children as a distraction, at least at times, and children often look at their parents in terms of what they will or will not do for them.

The I-Thou relationship occurs when two people encounter each other in a radically open and mutual way, when each person is present and available to the other, and when the other is an end in himself or herself rather than a means to an end. Such an encounter arises through grace in the sense that it cannot be willed, controlled, or predicted. However, this does not mean it happens involuntarily.[15] According to Buber, the moment of the encounter is a simultaneity of choosing and being chosen. Moreover, we have to be free from the force of fear and habit to respond genuinely and creatively to the new and unique in a particular situation.[16] Another way of speaking of freedom is to say that the person is self-possessed in the most basic sense of being with himself or herself.

It would be just as true to say that in responding to the unexpected in the situation, one finds a new measure of freedom. The presence of the other solicits a responsiveness and openness from the self, as is clearly the case for Carmen. Furthermore, the situation in which one sees the other "as if for the first time" is one that reveals who the other is in a much deeper way. In so doing it renders inaccessible, irrelevant, or at least significantly incomplete previously taken-for-granted or habitual ways of interacting with and perceiving this person.

Thus, it becomes understandable why in many instances we do not really allow ourselves to see, or to respond to, the full implications of what we start to become aware of in another person. In this story, we sense Carmen's initial hesitation and even pain at hearing what her son had to say to her because of the change and therefore the loss it brings with it. Yet, as she nonetheless allowed herself to hear what was being said and to affirm Davey as the person he was becoming, her love for him grew. She realized that Davey was not just her son but a separate human being who had thoughts and a perspective of his own and desires that might well conflict with those of his mother.

This coming to appreciate the other as having his or her own point of view and thoughts is part of all the descriptions I have looked at even while it occurs in a wide variety of circumstances. It is very much in evidence in a description written by a college student, Sheila, who told of her developing friendship with a fellow student, Sr. Lois, who lived in the same dormitory as she did. The growth in their relationship came when Sheila had a serious conflict with a teacher who was also a nun. Fearing that Lois would side with a fellow nun, Sheila did not tell her about the trouble she was having with this teacher. Finally, however, after a particularly distressing confrontation with her teacher, she took a "leap of faith" and told Lois about the conflict. Her friend did not respond as Sheila feared that she might:

> The sermon didn't come, the defending of the other didn't come, but a sincere attempt to understand and sympathize with me. We began to tell each other about school and this particular teacher. She told me her feelings about the situation, and I accepted them as from her and not as influenced by the convent.

Subsequently, Sheila went on to talk to Lois about an issue that previously she would not have discussed openly with any nun, showing how much she had come to trust Lois:

> I told her things about my relationship with my boyfriend that I would never have discussed with her before. I didn't get a sermon on values, relationships or love, but a person who listened and tried to help. I just knew these were her own opinions, something she had always believed. I wasn't seeing her anymore as a "nun" but as herself.

The response of the other was not what she feared and was more than she had hoped for. And in the process of being surprised she came to a new appreciation of the personhood of the other who was no longer subsumed under the category "nun."

2. Participation in the Perspective of the Other

"Seeing the other" is more than just a matter of paying attention to the person or of being surprised as a result of discovering something new about him or her. It is much more fundamental than being surprised to learn, for example, that our friend, whom we know to be an avid chess player, is also a keen swimmer. The notion of wonder points to the nature of that by which we are surprised. The basic question I asked about "seeing the other more as a person in his or her own right" implies a stepping beyond the boundaries of our self-interests and self-preoccupation. But what do we step into when we step beyond? Statements such as "we gain access to the point of view of the other" or we "come to empathize with the other" seem only partly adequate.

A passage in the French philosopher and child psychologist Maurice Merleau-Ponty's last work, *The Visible and the Invisible*, addresses this question in a remarkable way.[17] One of the issues he is concerned with is how one can make sense out of the fact that our existence is radically intersubjective (i.e., fundamentally connected with the lives of others), even while we also experience the world from our own perspective. In the following excerpt, he approaches this question by means of an intriguing account of how our relationship with someone we tend to take for granted may be disrupted:

> Here is this well-known countenance, these modulations of voice, whose style is as familiar to me as myself. Perhaps in many moments of my life the other for me is reduced to this spectacle, which can be a charm. But should the voice alter, should the unwonted appear on the score of the dialogue, or, on the contrary, should a response respond too well to what I thought without really having said it—and suddenly there breaks forth the evidence that yonder also, minute by minute, life is being lived: somewhere behind those eyes, behind those gestures, or rather before them, or against coming from what I know not what double ground of space, another private world shows through, through the fabric of my own, and for a moment I live in it. I am no more than the respondent for the interpellation that is made to me. To be sure the least recovery of attention persuades me that this other who invades me is made only of my own substance; how could I conceive, precisely as *his, his* colors, *his* pain, *his* world, except in accordance with the colors I see, the pains I have had, the world wherein I live? But at least my private world has ceased to be mine alone; it is now the instrument which another plays, the dimension of a generalized life which is grafted unto my own.
>
> . . . It is in the world that we communicate, through what, in our life, is articulate. It is from this lawn before me that I think I catch sight of the impact of the green on the vision of another, it is through the music that I enter into his musical emotion, it is the thing itself that opens unto me the access to the private world.[18]

This passage captures the quality both of wonder and of familiarity that is at the core of our ongoing experience of our fellow human beings. As in this example, we may be having a casual conversation with a friend, not paying much attention either to what we are saying or to his reaction. Then he responds, making it evident that he has been listening carefully, perhaps even understanding more of what we have been saying than we ourselves have. Now we are aware that this person is really *there*, looking at us. At that moment, Merleau-Ponty suggests, we live in this person's private world. Of particular interest here is that in this example, the world, the thing itself to which the other is attending, is ourselves. There is the uncanny feeling of being exposed as well as sensing the perspective of the friend as something quite tangible, something directly sensed, even if it is not visible. We are aware, for a moment, of being both the perceived and the perceiver, as we share, in some measure, in the perspective of the other who listens to and responds to us. The notion of participation—of sharing in a world rather than just having access to another's point of view—is central to Merleau-Ponty's analysis. This is, as we have already seen, far more than a case of detached awareness or an act of "accurate cognitive appraisal." What Merleau-Ponty is concerned with is altogether different: the other person's world compellingly unfolds for us, touching us emotionally and existentially.

The words of Vanessa, a woman who attended her younger brother Sonny's high school wrestling match because she wanted to understand something of his obsession with the sport, convey how such an unfolding can touch a person at a deeply personal level. Vanessa had been watching her brother diet for a number of weeks so that he could remain within a specific weight class on his wrestling team.

> It got to the point where I really began to despise wrestling because who can eat and enjoy something good when you have to watch your brother eat a single, solitary grapefruit for dinner because he's still a pound overweight? I decided then that I had to go to a wrestling match and see what it was all about. Since I didn't know what to expect I was all too ready to go, have a good laugh, and come home to tell Sonny that he was crazy!

Then she describes the wrestling match:

> After seeing the first three boys wrestle, I began to understand. Wrestling is a team sport, but first it had to be an individual sport for each guy is on his own. It's just him, his opponent, and the official on the floor and all the spectators' eyes are on you.
>
> I now began to see Sonny as a person in his own right. He wasn't just my brother, the son of Mr. and Mrs. Jackson, he was one, a part of others giving all that can be given out of them. Trying not to match muscle with muscle; but using every ounce of wit and strength in order to win for the team.
>
> As I watched Sonny wrestle, I saw him as a new person. Moreover, he lost and for the first time I saw him cry. Then I realized what it meant and

I felt bad because there was nothing I or anybody else could do. At this match I saw my brother become a person in his own right with disappointments to bear.

According to the German philosopher Max Scheler, the deepest level of recognition of another comes to the person who loves him or her.[19] Similarly, the American philosopher Maurice Natanson has suggested that it is only through friendship or love that one can be open to the other in such a way that the person "is recovered in his full integrity."[20] He contrasts this openness to the other with a much more distant attitude toward the person where he or she is seen as that individual, "out there." This element of friendship, affection, or love was evident in the previous four descriptions: from Denise, of her husband Gary (in Introduction); from Mary, of her mother; from Carmen, of her son Davey; and from Sheila, of her friend Sr. Lois. It was evident, though more implicitly, even in Rachel's response to Wayne.

Vanessa's account is reminiscent of Denise's description of her discovery that Gary has a life beyond his relationship with her. By watching him conduct a choir performance, she has gained access to a significant aspect of his world. Similarly, Sonny's sister Vanessa has gone out of her way to watch him wrestle and thereby understands not only what wrestling is all about but what it means for Sonny to participate in this sport, to give it his very best effort, and to lose.

3. The Separateness of the Other

The phrase "participation in the perspective of the other" might be taken to suggest that the two persons somehow "merge," especially given that there is a certain intimacy in the experience, and intimacy is often thought to be synonymous with "oneness." The descriptions I have collected do not support such a notion. If we participate in the perspective of the other, if we are there alongside of the other, we are still present as ourselves and the other as himself or herself. As the phenomenological psychologist Amedeo Giorgi has stated, one cannot have access to the viewpoint of the other in the same manner as that person.[21] No doubt Vanessa's viewpoint is directly related to Sonny's insofar she became sad as she recognized what it meant for him to lose. But she is sad at *his* loss, not her own. Emmanuel Levinas, a French philosopher whose work focuses on the interpersonal and its ethical implications, argues that the self-other relationship is irreversible.[22] Although I can come to appreciate the other's point of view, and even understand something of how the other looks at me, I cannot step outside of myself, viewing myself as the other does; nor can I experience the other as if I were him or her.

One of the most profound aspects of our becoming aware of, and present to, others is that we recognize them as a "source of meaning."[23] We recognize how they, through their very existence, bring a world into being. How remarkable it is that an intimate relationship opens up a new world for us. We learn that others attend to and care about aspects of reality different from those we do, just as they view the things around us differently than we do. This opening up to a new world is implicit in Davey's mother's statement that "Davey had a right to react in his own way." It is just as evident in Vanessa's realization that her brother lived in a world of wrestling, a world that she began to enter, through him, as a concerned observer.

The issue of separateness deserves further discussion. On the basis of our own experience, we know about the wondrous as well as the commonplace, and sometimes troubling, aspects of our relationships with those who are important to us. There are those moments when we feel very close to another person, and yet, before we know it, the other slips away and a distance suddenly opens up between us. In retrospect, we may believe that in the moment of intimacy, we lost our "selves," when, in reality, all we lost was our self-consciousness and self-preoccupation. In addition, however fully we appear to know or to appreciate the other, he or she always exceeds the grasp of our knowing and eludes the bond forged in special moments of understanding. We want those moments to reoccur, and they may, but not at the time and in the form that we expect and, clearly, not at our bidding.

Because moments of closeness and of deep contact are not lasting, and because separateness may often seem to be more like separation, some writers have taken the position that we are fundamentally alone.[24] Others, in contrast, have attempted to discern how separation can be overcome as if it were an unfortunate and avoidable part of life. Indeed, there are all sorts of gaps and barriers between people—misunderstandings, stereotypes, fears, and suspicions—that may be partly overcome. But the intrinsic distance among us also serves as a connection and draws us together. Buber has argued that what distinguishes human beings from animals is our capacity for setting things at a distance. It is the reality of this distance that allows us to enter into a relationship with what we experience as being independent and separate from us. Setting things at a distance and entering into a relationship constitute for Buber what he calls the principle of two-fold movement of human life.[25] Genuine relationship presupposes this realization of the separateness of the other. As Buber expresses it, "Genuine conversation, and therefore every actual fulfillment of relation between men [sic], means acceptance of otherness."[26]

In a similar vein, Levinas writes of the other as the stranger and the absolute other. He wants to emphasize that we can neither possess other people—however much we may try—nor fully understand them. Others are in some way elusive even when they are fully present to us and we to them; they reveal themselves to us and yet they are not transparent. The world in which we live is largely a shared world, but each of us approaches it in a distinct way. But this is part of what makes relationships possible and desirable. We cannot have genuine conversations with ourselves; instead, the call of relationship is precisely a call for us to move beyond ourselves.

4. Awakening of the Self

The recognition of the other has two sides. There is one's new awareness of the other person, and there is, at the same time and inseparable from it (however much in the background), a change in oneself. To see another "as if for the first time" is not only to experience this other person's presence but also to become psychologically and spiritually available to the situation as a whole. In these moments, we are far more receptive and responsive to the other than we are ordinarily. In contrast to such states of awareness, it seems as if we had previously been sleep walking. As I said previously, ordinarily we are restricted by long-standing but barely noticeable habits of action and perception. When Vanessa decided to attend the wrestling match, she was moved by a desire to make sense out of her brother's intense interest in wrestling. She was willing to move beyond her previous conception of who he was, and, implicitly, she was willing to be changed by what she learned. Similarly, Davey's mother, however much she was taken aback by her son's refusal to kiss her in public, genuinely wanted to know what he was thinking. Even if the other person is unaware of our presence—Gary was directing the choir, and Sonny was focusing on his opponent—we are responsive. That is, we are moved by what we see, and we realize that our previous conception of the other was largely false or inadequate. When we are receptive in this way, we are responding from a deep level within ourselves—a level that is heartfelt and that involves our most basic values and concerns. In this sense we are awakened not only to the other but also to ourselves and what ultimately matters the most to us.

Receptivity is often equated with passivity, with simply registering or receiving what is already there.[27] But one cannot do justice to these profound interpersonal experiences if one thinks of them in terms of a mere registering of some reality about the other that was previously just overlooked. To be receptive as a human being is also to be responsive and creative. Regrettably, the notion of "creation" carries the connotation of a subjective and internal activity in which one simply "makes things up." Thus, "creative bookkeeping" is another term for fraud. But this is by no means the only or the most authoritative meaning of the word. Buber speaks of "imagining the real," by which he means an extending of oneself toward the other with the aim of understanding what the other person at this moment is thinking, perceiving, and experiencing.[28] In the scenarios we have examined so far, as well as in the ones that follow, the receptive and the creative are aspects of the same experience. In being receptive to her son, Carmen moved creatively. As she responded from the depths of her own thoughts and feelings, something new was brought into the world in terms of the nature and quality of their relationship.[29] Thus, we can see that the awakening of the self to encounter or embrace more of the being of the other person is indeed a movement of creativity. In being receptive and responsive, the self changes, the image of the other alters, and the relationship changes in ways that were unanticipated.

Linda, a psychotherapist, gave a particularly striking description of this sense of awakening when she told the story of how an intense conflict with a colleague was resolved. She had instantly taken a dislike to Heather who was hired at the social service agency where Linda had been employed for several years. On the first day that they met, Linda got a strongly unfavorable impression of her new colleague. Heather came across as an aggressive, manipulative, and condescending person who gave Linda very little room to be herself. For example, on one occasion, Heather came into Linda's office and described how she had cleverly elicited the kind of response she wanted from a client. Heather spoke so quickly and insistently that Linda felt she was no more than an audience. In fact, Linda was so taken aback by what she heard that she found herself unable to express her dismay at the way her colleague had acted. This incident left Linda with an acute sense of discomfort. Afterward she felt as if her office were closing in on her. During our interview, she said to me, "The only response I knew was, 'Oh, my God, get her away from me, get her out, it is impossible, she is impossible.'"

The turning point in their relationship came several months after they first met. Linda walked into Heather's office and was astonished to find her crying and, in contrast to her usual manner, "not up to anything." Linda's first impulse was to leave, assuming that Heather would not want to be seen in such a vulnerable state. Nonetheless, she stayed to ask Heather what was wrong and found out that Cathie, a close friend of hers, had been diagnosed with a terminal illness. This friend of Heather was also a woman whom Linda admired and respected. Learning about this woman's illness saddened and troubled Linda. "But more than anything else I was kind of struck by the look on Heather's face. The pain and the vulnerability, the weakness, just being a person with human feelings and human pain." For the first time, Linda felt able to relate to Heather, no longer suspecting that she was "up to something." Throughout the afternoon, Linda was supportive of Heather, who, in turn, accepted this support, and they were able to share personal feelings in a way that was in sharp contrast to how they had previously related.

This day of shared grief was a milestone in their relationship. "Seeing the other as if for the first time" has been described as a landmark in all of the descriptions I have collected. We let barriers of antagonism, fear, envy, resentment, suspicion, or apparent indifference collapse, momentarily at least, as we respond, creatively and deeply, to the surprising way in which the other reveals himself or herself. On these occasions we realize that our previous conception of that person was unfair, incomplete, or in some other way missing the mark. It is not just that the content of our conception changes, but that we connect with and come into the presence of the other person as a genuine other, as much more than just a projection of our own wishes or fears. Denise, in coming to an awareness of the separateness of her husband, moved toward a more mature understanding of her relationship with him; Carmen, in affirming the autonomy and individuality of her son, also moved forward in terms of her own development as a person.

I would note, however, that these landmarks can also be troubling or painful in nature. One man I interviewed spoke of becoming aware, after he saw his wife in intense physical pain, of how he had taken her for granted. He had assumed that his wife would always be with him, and it had never occurred to him that she would ever experience physical pain. "I was worried, I was confused, I didn't know what to do. I was disoriented, that situation had never come up before." When she was in pain, he feared that her life might be in danger, and he became acutely aware of her mortality.

After breakthroughs such as these, we go through a process of reevaluating the past, wondering how we could have failed to see what we assume must have been there all the time. This reevaluation typically leads to asking some troubling questions about oneself and one's stance toward the other in the midst of what is otherwise a very moving experience. We may conclude that we were too insensitive, immature, or self-centered to really attend to the other person. These reflections, however justified, may distract us from acknowledging the particular circumstances that allowed us to come into the presence of the other.

5. A Horizon of Hope

Regardless of the presence of such self-questioning, the experience is one that has hope at its core. For some, it is the unexpected realization of an already existing desire to be closer to the other person. This was clearly the case with Mary and her mother. For others the experience brings with it a genuine appreciation of the other person, and thereby the desire for further closeness arises. This was the case for Linda as she shared Heather's grief.

But to fully appreciate what hope means in this context requires further reflection. The philosopher Gabriel Marcel, who has written about hope and its intimate relationship to despair, has given us an especially thoughtful and rich interpretation of this phenomenon. He has characterized hope as arising from participation in life and openness to its ambiguity, in distinction to optimism or pessimism, both of which involve a distancing from the ongoing flow of experience.[30]

Marcel's approach to understanding hope is existential in that he appeals to the experience of the reader, to his or her already implicit familiarity with the core of this phenomenon. Rather than starting with a definition, which could then be supported logically, he uses concrete examples to arrive at a clarification of the essential nature of the phenomenon. He believes that we need to look at hope and despair together because they imply each other (despair literally means to be without hope) and because the one becomes visible in contrast to the other just as light is always juxtaposed with darkness. For example, someone who has been in despair for long time may only become explicitly aware of how hopeless she was feeling once she finds new meaning in life.

Life is such that despair and hope are both givens. In this regard, the theologian Walter Davis has argued that "optimism, not despair, is the enemy of hope."[31] Davis believes that optimism blinds us to the dark side of reality, and thus is a form of denial. This is a point Marcel similarly elaborates on when he suggests that optimists and pessimists are equally spectators to the flow of life. While disconnected from the depth of being and the events around them, they take on an observer-like attitude of knowing "what really is the case." "In contrast," Marcel writes, "he who hopes is engaged in a process."[32] That is, the person who hopes is genuinely open to the ambiguity of life.

In the case of Linda's unfolding relationship with Heather, we were first told how impossible or hopeless the situation between them was—Linda felt she could not be herself in Heather's presence. Subsequently, Linda was so affected by Heather's vulnerability that she became "more herself."

To experience the other as if for the first time entails, then, a movement of the self toward openness and receptivity, with hope as background or horizon. In Linda's description, this sense of becoming connected with the flow of life is particularly apparent. Before Linda came to share Heather's grief, she had felt extremely guarded and uncomfortable because of her antagonistic relationship with Heather. In what follows she speaks eloquently of a meeting she had shortly after her encounter with Heather, describing her new sense of freedom and connection:

> I guess I felt liberated, and I ended up by the end of that session with the client feeling much more in touch with what living and dying are all about, and the temporary nature of human living. Somehow being really in touch with that through the tragic news of Cathie's illness, somehow enabled me to be more, I don't know . . . to be in tune in such a way that was helpful to the client.

If we think back to Mary's story about her mother, it is evident that the experience of understanding gives hope to both of them. Her mother no longer feels so isolated and Mary no longer dreads the moment when her mother becomes the "monster." Both have found that movement is possible, that things do not need to remain as they are, and that connection between human beings is more than something one merely wishes for. In van Kaam's analysis of "feeling really understood," the element of connection with others was an essential element. What should be added is that the listener who "really understands" is also touched by this meeting, by this coming together. As Arthur Egendorf points out, when someone speaks openly to us about something that is difficult or painful for them to reveal, this person is essentially telling us that we are trustworthy, and this is a way of honoring us."[33]

Let us take a moment to revisit Rachel's story. There the presence of hope as horizon may be less obvious. She speaks of the possibility she sees in Wayne, someone both she and her boyfriend, Wayne's brother, have worried about because he is "a strong man given to excess which could be controlled when he found what or who he was looking for." She adds, "The potential I thought I saw startled me." There is an ambiguity here: things might turn out well for Wayne, but not necessarily. There are no guarantees. But this incident and the conversation they have subsequently lead her to see Wayne as a friend rather than as her boyfriend's brother. There is movement here for her; there is an opening up, and in that opening up and the connection it brings, hope dwells quietly.

Marcel[34] has said that to despair of oneself is to despair of the other. The reverse also seems to be true. These descriptions point to the fact that to become hopeful about the other is to become hopeful about oneself, even as one is connecting with the other in an intimate way.

The Context for the Recognition

In reflecting on the stories I have collected, I have distinguished four specific types of contexts in which we may come to see the other "as if for the first time." The distinction among these types is not absolute by any means, however.

In the first type, the initiative of the other person leads to a change in our perception of him or her. That is, it is what the other person does that causes us to "wake up." For example, we remember that Vanessa was taken aback when her son Davey told her that he was too old to be kissed in public. Similarly, Professor Openshaw's clerk, Mr. Berridge, dramatically demonstrated that his employer's view of him was woefully inadequate. In each case, it is not just that the other acts or approaches us, but that he or she surprises us in such a way that we become fully attentive, or, conversely, that the other captures our attention in such a way that we are surprised.

Joan wrote about coming to a new understanding of her friend Elizabeth who was "the most gorgeous girl anyone could ever dream of" and who was also a model. One summer morning when she visited her friend, she found that the drapes were still drawn and Elizabeth looked anything but gorgeous as her makeup was streaked and her eyes were red and puffy. Elizabeth explained that she was so distraught because she thought she had finally found a man who appreciated her as a person rather than just valuing her good looks. She had showed this man the poetry she had written and had opened up to him in a way that was unusual for her. Then, at a dinner party the night before, this man had made fun of her poetry and starting telling the others there about some of the things she had told him. She was devastated. As Joan and Elizabeth continue to talk, Joan had the opportunity to read some of Elizabeth's poetry. She also realized, to her surprise, that Elizabeth cared what her reactions were. In summarizing what happened that day, Joan wrote, "I've known her for so long but it seems that I just met her this summer."

The other person opens up to, approaches, or confronts us, and in that moment of surprise we are practically forced to take a second look at him or her, seeing this person in a new way. Nonetheless, it requires interest and willingness on our part to listen, to see the other in a new way, for anything really to happen.

The second type of context is the mirror image of the first. We take the initiative in approaching the other person and, thereby, provide the opportunity for the other to reveal something of who he or she is. Sheila approached Sr. Lois to talk about the conflict she was having with another nun, her teacher. Lois' response did not fit with what Sheila feared a nun might say or think, and Sheila started to see more of who Lois herself was.

Within the context of a family, it might be very difficult or scary for a son or daughter to open up to a parent. One student wrote about what happened when she finally told her mother that she was struggling with her choice of vocation. Emily grew up in a family in which there was a long tradition of everyone becoming a physician or dentist, and so she felt that unless she followed this path she would be letting her family down. It did not occur to her until she reached her sophomore year in college that she might not enjoy the medical profession and that there were other possibilities for her. It was very difficult for Emily to speak to her mother about her developing interests in liberal arts because she was convinced her mother would be really disappointed in her. Eventually she did speak to her mother, and although at first her mother suggested that Emily was too young to make this decision, she gradually came to accept her daughter's new interests. Emily's own words capture her realization that these conversations with her mother involved a change in their relationship as well as in her.

> After I had broken through the barrier I found out that my mom was open to my feelings and that she became more aware of the fact that I was an individual, I felt that I could be more opened and unafraid of confronting her with certain problems and decisions that occurred . . . She told me her feelings on certain issues and in turn I told her my feelings; actually we opened up in communication for the first time. As a result I saw my mom in a different light because I was growing up.

A third possibility is that we gain a deeper appreciation and awareness of another when we see that person involved in a situation other than the one we share with him or her on a day-to-day basis. That is, we see the other person on his or her turf or in relation to a world that we do not share. When Vanessa attended the wrestling match at her brother's high school, she became aware of a dimension of Sonny's life she had not understood before. In so doing she came to see him in a new light. Similarly, Denise's experience of her husband Gary as someone who had a world beyond his relationship with her came when she saw him in his role of choir director.

The fourth context is one in which two people come to be fully present to each other as the reason for their being together changes in some important or even radical way. This change is illustrated by Heather and Linda's encounter. They came to share their concern for a dying friend, a context for being together unlike any they had shared before in their work as colleagues. When there is a history of long-standing or intense antagonism, or when two people have come to take each other for granted, it may well require an extraordinary event to bring about change, as was the case in this instance.

All too often family members do not really become attentive to each other until illness or some other disturbing event disrupts their habitual ways of interacting with and perceiving each other. But such a disruption by itself is not enough. It has to bring with it the possibility of a shared ground of concern and value. In the next chapter we will look at the experience of disillusionment where there is a disruption in a relationship without the emergence of a new basis for remaining connected.

Conclusion: The Heart of the Matter

I want to conclude by focusing on what is in common in these various types of contexts. Most fundamentally, the context is one in which the other person is fully present and engaged at a deeply personal level, as with situations that involve grief, concern, joy, or pain. The point of view of the other and his or her expression of feeling, whether it is directed toward ourselves or some facet of the immediate situation, strike a responsive and empathic chord in us. That is, the person's behavior and expression are seen by us as readily understandable and valued manifestations of the other's basic humanity. Further, the manner of the other's presence is unexpected in terms of our preconception of who this person is. The event or issue with which the other is engaged is one that speaks compellingly to our basic values and concerns, thereby calling forth from us a basic openness to him or her. Standing in the presence of death is one such event.

As we have already seen, Linda and Heather were brought together by the news of the imminent death of their mutual friend. The German philosopher Martin Heidegger has insisted that death is a basic horizon of human existence.[35] Death brings us to the realization that our time and the time of those we love are limited. In the face of this recognition of finitude and the urgency to which it gives rise, our inhibitions and reservations may dissolve. The Dutch psychiatrist J. H. van den Berg reminds us, "And the people we love, those few who are with us for a while, how could we love them if they did not grow and die? It is the light of death that makes them dear to us."[36]

However, this is not to say that it is only in the face of death that we open up to others or that others become visible to us. The stories narrated here suggest otherwise. Rather, as I said earlier, the key point is that we are responding from a deep level within ourselves as the situation in which we find ourselves awakens us to the reality of the living presence of the other person. It is also a situation that we experience as complete in itself. That is, we are genuinely present to a situation that makes sense to us. We have a genuine moment of understanding the other and being present to him or her whether or not that presence is reciprocated. Of course, insofar as some aspect of the situation calls us away or troubles us, our focus may be short-lived. The man who was confronted with his wife's mortality when he found her in intense pain quickly moved to find some way to bring help to her.

Our response to the other, whether expressed directly or not, is character-ized by an absence of reserve and self-consciousness. Think of how Mary started to cry when she understood how lonely her mother's life had been, or of Vanessa's sadness at seeing her brother lose his wrestling match once she understood how much it meant to him. This experience of epiphany involves a deep recognition of the other as other and, at the same time, it involves, an awakening of the self. It brings us face-to-face with human mortality—the vulnerability of our humanity—even as it is grounded in our greatest strength, that is, our capacity for caring and for loving.

CHAPTER 2

On Being Disillusioned by a Significant Other

Maturity, like all coming to oneself, requires a break with what has been inherited. We come to ourselves only in confrontation with the other.

Erwin Straus[1]

When we are genuinely surprised by someone we know well, the surprise is likely to become a milestone in the relationship with that person. In the previous chapter, we looked at the experience of coming to be present to the other "as if for the first time." In this experience, we discover in a compelling and concrete way that the personhood of the other far exceeds the image we had of him or her. A deepening of the relationship characteristically follows.

But, as is painfully obvious, what we learn about other people may be far from encouraging or positive. Disillusionment entails being surprised by the other in such a way that the relationship is undermined. Rather than uncovering a common ground of shared values and understandings that brings about a strengthening of our connection with the other person, the nature of the discovery is such that it brings into question the very foundation of the relationship. One's assumptions about the other person's positive qualities are shown to be illusory. In describing how she was disillusioned by the psychotherapist whom she had greatly admired and liked, Joan, one of the people I interviewed, said, "It was like the loss of that friend that hadn't really been there."

When we are disillusioned by someone close to us, we are not only profoundly disappointed in that person, but the very meaning and direction of our life and our relationship to the other simultaneously come into question. Most of us recall, all too well, whether from the perspective of the one who is disillusioned or the one who has failed to live up to someone else's expectations, times of deep disappointment and their aftermath. It is painful to recall those times when we have let down someone who looked up to us. Thoughts of "if I only had acted differently," or "if I could just make it up to this person," are accompanied by feelings of shame and regret that cut deeply into us. No longer being the one who is highly valued by the other, and seeing the cherished relationship fall apart, can be a devastating loss.

As we know, children rarely fulfill the high hopes that parents have for them just as parents fail to meet the idealized expectations of their children. Not only is disillusionment part of life but, as the quote from the psychiatrist Erwin Straus (at the beginning of the chapter) suggests, insofar as it is a confrontation with the reality of how another person is different from oneself, it may provide an impetus for a movement toward greater maturity. By the same token, however, if it is an experience for which we are ill-prepared, being disillusioned can have extremely painful or even tragic consequences.

The aftermath of disillusionment—how one attempts to cope with the sense of loss and betrayal—relates closely to a topic that my colleagues and I have been studying for a number of years, namely, the psychology of forgiveness.[2] As I will show in Chapter 3, forgiving the other, as well as oneself, may be part of the process of coming to terms with the loss of illusion.

Again, I started with first-hand descriptions, more than twenty-five written and two solicited through in-depth interviews. As was the case in the first chapter, the people who provided the stories have varied backgrounds, in terms of upbringing, ethnicity, and outlook. They range in age from twenty to fifty years; two-thirds were women. In each case I asked the person to tell about a time when he or she was disillusioned by a significant person in his or her life. When I had questions about the written descriptions, I met with the person subsequently to seek clarification.[3]

Seven of the descriptions involved being disillusioned by a father or stepfather; all of these were written by women. Other family members by whom the person was disillusioned included an uncle, a sister, a brother, and an older cousin. In nine cases, the significant other was a friend or best friend; the remainder involved mentors, a therapist, lovers, and colleagues.

The analysis of these descriptions was guided by the phenomenological ethos of being as faithful as possible to the meanings of the persons' experience (see Introduction). As part of this emphasis, I have made liberal use of the individuals' own words to present central aspects of the experiences, analogous to what I did in Chapter 1. After discussing the experience of disillusionment, I will relate it to the existing literature on this topic and conclude the chapter with a dialogue between this literature and the stories.

Disillusionment and Its Aftermath

The Ground for Disillusionment: The Idealized Other

The stories that follow describe the idealization of the other that preceded the disillusionment. They also give some sense of the history of the relationship, providing the context for such idealizations. As we will see, psychoanalytic writers assume that idealization is a necessary prerequisite for loss of illusion, and these descriptions, with one possible exception, support this assumption.

What does it mean to say that the other is idealized? And what is the nature of the relationship between oneself and the person one idealizes? These two aspects are intimately connected. There are many dimensions and layers to idealization, but fundamentally it means that the other is seen as larger than life, as having positive qualities, such as being powerful or caring, that make the person special. Frequently, these are qualities that we wish that we ourselves had. The stories that I have worked with demonstrate that the identity and hopes for the future of the person who idealizes are very much part of the relationship with the person who subsequently lets him or her down.

Ellen's story of her relationship with her father illustrates both of these points vividly. She had grown up believing her father was the "good" parent and her mother the "bad" one. Her mother was the disciplinarian, the one who said no, who frequently spoke to her in a denigrating way and left her with the feeling that she was a bad person. Her father, in contrast, was lenient and generous, continually affirming that Ellen was capable and worthwhile and providing support and structure that helped her to be successful and to feel that she was a good person with a promising future. She saw him as a strong person and was very much aware of his remarkable professional accomplishments. "My father," she remarked, "has always been very tied up in my view of myself."

For Benjamin, a man in his early thirties, the idealized other was a woman with whom he became intimately involved. She was very pretty, athletic, well educated, and spontaneous—he really enjoyed her company. They had long conversations together, and he believed that she was faithful, honest, and committed to him. She represented much that he admired and valued, and he thought he had a potential future with this woman who seemed to be so special.

Such a clear dependence of the meaning of one's own life on one's relationship to the significant other existed at an even more radical level for Jasmine. This was a woman who remembered how her father rescued her from drowning when she and her family were on a boat that capsized while they were escaping from a Southeast Asian country. Thinking back to this incident, which happened when she was six, this woman wrote, "My father was not only my hero, but he was the hero of several other people whom he saved." She continued to hear good things about her father from other family members and the larger community once the whole family had settled in the United States. And when her parents argued, as they often did, Jasmine blamed her mother for demanding too much of her father.

Another young woman wrote poignantly of what her relationship with her stepfather had meant to her. Although Margaret initially opposed her mother's marriage to David because she had hoped her parents would reunite, she gradually accepted him, and, she added, "I came to love him as a father." She was moved by the acceptance he and his family bestowed on her, her brother, and her mother. As a thirteen-year-old who was very nervous about beginning junior high school (and fitting in), having a step-father gave her a sense of belonging to a "real nuclear family." Her life changed for the better as she came to accept David as a dependable and supportive adult in her life. In her own words,

> I started allowing myself to be a kid again instead of an adult, and I began making new friends. David took me to dentist appointments, he met and interacted well with my friends, he taught us to work together as a family team, and he basically became the "man" of the household.

Richard, a man in his early twenties, reflected back on his adolescent admiration for a cousin who was two years older and had a "type of spell-binding power over" him. His cousin, he assumed, knew what to do and how to do it—he was someone who was on top of things. The cousin provided Richard with the brother he had never had as well as a model of what growing up meant.

A similar theme is sounded in a description from Rosemary, a woman who worked in a campus ministry program. During her high school years, Rosemary had greatly admired her church's youth ministers and enjoyed the activities they arranged for her and her peers. When she entered college, she had "magical images" of campus ministry and the people who worked there, based on her positive experience in high school. In particular, she wrote admiringly of an Episcopal priest who had been hired to work in this program: "She had done everything . . . she was a priest and I always wanted to be a priest and cannot because I am Catholic."

In contrast, Jeannine, a woman in her thirties who spoke about her disillusionment with Doris, her younger sister, said little that explicitly pointed to idealization. The story is particularly interesting because it both overlaps with and is different from the other accounts. It is what one might call a "borderline" or anomalous case, allowing us to see more clearly the features of the stories that fit unambiguously into the pattern of idealization followed by disillusionment. For this reason it merits careful attention.

Growing up, Jeannine and Doris had little in common. However, when Jeannine moved back to her hometown after graduating from college, she made a concerted effort to reconnect with Doris, who was now pregnant, and her husband. She wanted to develop a sense of family, something that had been absent in her life previously. Jeannine was included in family activities, they celebrated holidays together, and they did things to help each other. Nonetheless, fundamental differences remained between the two sisters. While Jeannine lived an independent life and believed strongly in confronting and working through life's difficulties, Doris was "more of a homemaker" kind of person who had not learned to drive and who pushed away anything that challenged her view of the world.

The birth of Jeannine's niece added to the sense of family but raised another issue of conflict—the possibility of homeschooling. Jeannine was convinced that homeschooling would mean that her niece would lose the possibility of developing socially through interaction with peers. Mindful of how much pain there had been in her own growing up because of her parents' limitations and of how going to school had helped her, Jeannine did her best to persuade Doris not to homeschool her daughter. The hope that Doris would make the decision that Jeannine believed was critical for her niece's well-being set the stage for her disillusionment with her sister. She feared a repetition of a family pattern even as she wished for a new direction that would sustain the tentative sense of belonging that she was enjoying with her sister and her family. So, although Jeannine did not "idealize" her sister, it was nonetheless obvious that her own future—and in a more subtle way, her sense of self—was significantly caught up with her sister's actions.

In these examples, as well as in the other stories that I have collected, the idealized other is seen as being powerful by virtue of his or her special qualities (or position, in the case of Doris) and, typically, as more powerful or attractive than oneself. Yet our reliance on the other whom we idealize is not regarded, at least consciously, as a problem insofar as the other is viewed as a benign figure who can be counted on to act in a positive manner toward us. For Jeannine, who lacked confidence that her sister would make the "right" decision, this dependence on the other was more obviously problematic.

Experiencing Disillusionment

One of the striking features of disillusionment is that it exposes the foundation that made it possible, while at the same time undermining or eroding that foundation. Typically, this erosion is a gradual process: one incident after another creates doubts about the other and then a particularly jolting event brings about an irreversible shift in the now disillusioned person's basic outlook.

Occasionally, the loss of illusion takes the form of what, at first sight, appears to be a sudden and dramatic moment of recognition. In either case, disillusionment highlights previously taken-for-granted patterns of beliefs, assumptions, and conduct at the same time that it exposes them as illusory. The "seeing of the other" in disillusionment, then, is primarily that of seeing who the other is *not*, rather than a recognition of the perspective and world of the other. One might say it is first of all a negative seeing, a repudiation, rather than an embracing of the other.

As in the experience of forgiveness, disillusionment simultaneously changes both past and future, but with a critical difference. In forgiveness there is resolution, with the future opening up as the person is reconciled to the past. In disillusionment there is conflict and contradiction, which brings into question or makes a mockery of the past at the same time that it closes off the future.

Furthermore, the shift that occurs has reverberations that reach beyond this particular relationship. Insofar as the other was an important part of our world, we start to look at that world and ourselves in a new way, and what we see is disorienting and even shocking. Benjamin, who eventually was completely disillusioned by the woman he had so admired when she left him to be with her ex-boyfriend, was acutely aware of these broader implications. In referring to the crisis he had been through, he exclaimed that he had been "betrayed by a dream." His entire vision of what constituted the "good life" collapsed as this particular relationship ended, and his hopefulness, however precarious, gave way to despair. Here too, as with the stories presented in Chapter 1, there is an "awakening of the self," but within this context one awakens to self-doubt, confusion, and loss of direction.

As with the experience of "seeing the other," this realization is not sought out. Indeed, as the above discussion suggests, disillusionment is a revelation one tries to avoid. The avoidance, while typically not fully conscious, can take a variety of forms. We can deny (to ourselves and others) that there is any problem at all even while feeling vaguely uneasy and apprehensive. Or we may minimize the implications of evidence that we do take note of explicitly. Thus, we might argue, our friend who has sworn off drinking is taking a risk hanging around with his former drinking buddies but is doing so to encourage them to consider sobriety. When disillusionment does come, it is characterized by surprise, its tone one of dismay rather than wonder.

Turning to the completion of Margaret's story about her stepfather David, we see how painful this sort of collapse can be. Over time, there were an increasing number of arguments between Margaret's mother and David, at the end of which he would walk out and be gone for hours. Yet both of these adults kept reassuring Margaret and her brother that fights did not mean the end of a relationship. Eventually, though, as other problems developed, David left Margaret's mother. Shortly after, he met with both of the children and assured them that he wanted to continue to be part of their lives and that they could still count on him. They never saw him again.

Margaret wrote, "I was mortified, the family I thought I had was only an illusion. . . . I felt like I had been physically beaten and I didn't know why." Here we get a glimpse of the shame, the self-doubt, the loss of a sense of belonging—the overall devastation that may accompany disillusionment, especially for a young person. After David left, Margaret's mother became financially insolvent and withdrew psychologically. The children expressed their grief and anger by acting out—Margaret through drugs and sex, and her brother through violence.

Ellen, who had idealized her father as the "good" parent, came to a new perspective on her family after she moved away from home at the age of twenty. She gradually realized that her father had set her mother up to be the "heavy," or villain, in the family, that he was quite passive and had rarely intervened when her mother was abusive. All of this, she noted, "didn't fit in with my idea of his being this strong person who kept everything together, who was incredibly wise." This realization raised doubts for her about the meaning of her father's validation of her own positive qualities.

One woman wrote of how her view of her father changed dramatically at the age of eight when her mother died during a hospitalization. Up to that point, "I thought my father was the smartest, bravest, most loving man in the world." But her father withheld the news of her mother's death from her and her siblings for a week, and then she learned of it through a cousin. Subsequently, he refused to discuss this tragic loss with her. In attempting to capture something of the feeling of the devastation she felt at this young age, she wrote, "Where was the father who fixed my broken dolls or braved impossible odds? My father had shrunk from a giant to an average man." The words of another woman, whose father had failed to stand up to an aunt who had been very abusive to her, similarly summarized something of the core of all of these women's disillusionment with their fathers or stepfathers: "Where is the man who was supposed to protect and defend us?"

And, I would add, what are the questions this failure or absence of the father raises about one's taken-for-granted self-understanding? When Jasmine heard that her father who had saved her from drowning had left his family in order to live with his long-term mistress, she became very bitter: "I felt betrayed and I was angry at myself for admiring and respecting my father while all those years he was lying to all of us by pretending to be a just man." She wondered if he had ever really cared for his family. Like so many others, Jasmine came to believe that she had been fooled by the person she admired and that this meant that she had been foolish in trusting someone so much. Later, her father returned home with a terminal illness. Jasmine did not cry when he died.

In the last year of high school, Richard discovered that the cousin whom he so admired was having an affair with his, Richard's, girlfriend. "This was a tragic day for me. I remember having a real sense of loss, a sense that something wonderful had come to an end, and that there was no means to get it back," he said. This statement illustrates a core dimension of disillusionment—it brings with it radical discontinuity as it shatters people's hopes for their future.

Jeannine also vividly described how sudden and irreversible such a shattering can be. When her niece was five years old, Jeannine, her niece, and her sister Doris were returning from a day of relaxation and fun in the country. As they stopped at a red light, Doris told Jeannine that she was going to homeschool her daughter. Her reaction was one of shock and disbelief, and yet there was also the realization, "Oh, my God, she really is telling me what she is going to do." Jeannine added, "Being trapped in the car at the stoplight was a metaphor for how and why this got through to me so deeply. I definitely saw her without my own hopes, wishes, fantasies projected on her or out of my own head. And I think I just really saw who she was." But this clarity of vision was above all a seeing of who her sister was *not*, as opposed to coming to an appreciation of her point of view. The shattering was not of a positive or idealized image of her sister, but of a faint and yet strongly held wish that Doris might somehow be persuaded to "do the right thing."

In retrospect, Jeannine could see that her disillusionment was very predictable, and she even joked about her naive belief that Doris would listen to her. In that moment in the car, Jeannine's desire to be part of a family was shown, beyond any doubt, to be at odds with her insistence that her family be different from what it had been in the past. Her hope that "the next generation would suffer less" was destroyed, as was her relationship with her sister as a sister. No longer did she seek out her sister as someone to have family time with. Rather, she continued to have some contact with Doris so that she could remain connected with her niece.

The letdown that Rosemary experienced in her relationship with the Episcopal priest whom she admired came gradually. At first, when this woman showed little interest in her ideas or the program with which she was involved, Rosemary explained to herself that this attitude was due to the priest's being new to her position and her enthusiasm for other programs. Over time, Rosemary awoke to the painful reality that this priest had little sensitivity to those with whom she worked, and that she was both single-minded and authoritarian in promoting her own ideas. Rosemary's initial sense of hope gave way to a strong disappointment. Sometimes, gaining clarity plunges us into darkness.

Aftermath

The experience of disillusionment presents a serious challenge to the coherence and meaningfulness of our existence. Finding new meaning and direction in one's life can be extremely difficult. However much the loss of illusion also involves an awakening, the life to which one awakens may initially seem untenable. This is evident in the case of Benjamin, whose dream of the good life fell apart as his girlfriend returned to a former relationship. Even though he was determined not to become embittered, as had others he knew, he was not sure at first whether he would "make it." It was not until a year later that he finally felt he was on solid ground. During this period he had made fundamental changes in his life and sought help from a number of people, including a priest.

The process of coming to terms with disillusionment includes turning to others for direction and support. The specific ways in which others can be of help further cast light on the meaning of disillusionment. That is, disillusionment brings with it disturbing questions about our personal worth and makes us wonder whether we have been deluded about what has been real in our life. Both of these issues can be addressed in a supportive relationship where another person confirms our value as a person and helps us to make sense out of our confusion, pain, and disappointment.

At the beginning of this chapter, I referred to a woman who was disillusioned by her therapist. Joan had been troubled because her therapist had acted in ways that she regarded as unprofessional. She ended therapy when her therapist insisted that Joan see her husband, also a therapist, rather than herself. During their last meeting, her therapist added insult to injury by charging that Joan's "character defects" had prevented her from benefiting from therapy. In response, Joan sought out a psychiatrist as a kind of "higher power." The fact that he was appalled by how her former therapist had acted was deeply validating for Joan and helped her to move from self-doubt to anger. Partly because of subsequent work with yet another therapist who listened empathically without taking sides, she was able to come to terms with what had happened. "After the anger and hurt subsided," she wrote, "I had to eventually step back and view my therapist and the situation from the perspective that she was on her own journey with her own experiences guiding her and that although her actions had hurt me deeply, the hurt was not purposeful or intentional." In a word, she had attempted to imagine what the other's perspective might have been and, through that process, to move toward forgiving her.

Two years after Jasmine's father died, she still had not discussed what had happened with anyone. She wrote that this period was so incredibly painful for her that she became preoccupied with thoughts of ending her life, not surprisingly so, given that the father, who had saved her from drowning and had been the hero of her life, had also betrayed her. Eventually, she went on a religious retreat and confided in the retreat director. This man was a priest who reminded her of her father—"the loving, gentle man I always knew before my anger and hurt had overshadowed his goodness." This was the beginning of a process of healing, which gave her a sense of inner peace and enabled a forgiving of her father, and thus allowed her the sense that he was still part of her life. Her conversations with the priest gave her back her life.

None of this is to suggest that everything was resolved and no pain or distress remained. For this woman or for any of us, it is impossible to return to the world as it was before disillusionment. But, as the psychologist Bernd Jager has argued in his discussion of how children develop autonomy in relation to their families, the critical issue is whether we can come to express and understand how we are affected and wounded by others, especially members of our families.[4] Part of becoming a mature adult, in other words, is leaving behind a state of "innocence" and coming to terms with the ways in which we have been hurt by others or have hurt them. Rollo May has eloquently described this movement from what he calls "pseudoinnocence" to a position of personal power in his book *Power and Innocence*.[5]

By pseudoinnocence, May means a refusal to acknowledge unpleasant realities and a holding on to a childish outlook on ourselves and others, one that ignores the darker sides of own nature and those of others. In contrast, when we move forward and become self-assertive, we are ready to meet resistance, "we make greater effort, we give power to our stance, making clear what we are and what we believe. We state it now against opposition."[6]

The following story illustrates this process. After being let down by a close friend, during a time when her need for support was great, one woman wrote,

> I've done a lot of growing in the past year, especially during the last six months. Because of a bitter experience, I view a friend differently, more realistically. This in the long run, really has worked out for the better. I also view myself differently. I know now I can depend on myself. This is a good feeling.

Who the validating, supportive other is varies from story to story. For Ellen it was her boyfriend and friends, people with whom she could freely discuss her changing evaluations of her parents. For Benjamin, it was a priest, a spiritual director, and several friends. For Rosemary, who was disappointed in the woman priest, support and understanding came from another person in the campus ministry. Jeannine, as part of her own therapy, is working on coming to terms with what happened with her sister. She also implied that her interview with me was a means toward getting a new perspective on what had happened.

To say that disillusionment presents a serious challenge to the meaningfulness of our lives is also to acknowledge that it brings into question one's sense of self—in many ways and at various levels. Although one can no longer be quite certain who one was before, one may also be catapulted forward. In this respect, Richard's characterization of his transformation captures a common theme among those who were able to make some headway with their disillusionment. There was still a lot of pain, he said, in thinking back to how he was disillusioned by his cousin, but "maybe more importantly I have memories of . . . a change within myself, a change where I no longer had to view the world through another's eyes."

Rosemary gave voice to the same theme, emphasizing that what she had gained from the experience was to "really trust the way I felt about things" and "a better sense of myself and how I work with different people." One woman who was hurt by the meanness of a jealous friend said that for her the lesson was that she should pay attention to her gut feeling that someone might not be good for her. In relation to disillusionment by parents, this task of clarifying self-definition and becoming more independent is especially difficult and complex. Ellen spoke of her struggle to separate from her parents and of how she sometimes felt unsure of who she was. Yet forgiving her father helped her to reclaim an image of herself as a capable person. That is, by forgiving him, she was acknowledging that he was a fellow human being as well as her father and letting go of the requirement that he live up to her ideal of who he should be. In so doing, she was affirming that she was a person who could move on with her life even though she had been deeply disappointed.

With this issue of change in the self in mind, I want to turn to a last question in regard to the outcome of disillusionment, namely, how does such an experience change one's view of, and relationship to, human limitations and fallibility? A young man wrote of his increasing disappointment in his friend and coworker who did not appear to take the work they did together seriously. As he found out more about the friend's history, relationship with parents, and point of view, he could eventually say, "I learned the difference between understanding people's differences, and 'allowing' them to be different from me." One way to view others is through the judgment that they are failing to meet our expectations; another way is to see their behavior as expressive of who they are rather than just as weaknesses or shortcomings. Allowing for difference between oneself and others, or allowing others to be different from how one expects them to be, is an even larger challenge where one's family is concerned. The "allowing," insofar as it occurs, is apt to be fraught with hesitation or ambivalence; the reconciliation, inasmuch as it comes about, may have an undertone of disappointment. This was evident for one woman whose father failed to protect her against a domineering aunt: "Now I am much more realistic, I see him as a human being, and an imperfect one at that." Jeannine, who was appalled by her sister's apparent failure to respond to her young daughter's needs, struggled to come to some acceptance of who her sister was and was not.

We can also think back to Joan whose movement toward an acceptance of her therapist's limitations took a number of years of intensive work. That work included coming to more of an understanding, on the basis of her own experience as a parent, of what it was like for her parents to be parents. She started to recognize how, in spite of their best efforts, her parents had made serious mistakes and that, in this respect, she was not any different. There was a similar shift for Rosemary, who gradually realized that "priests are like everyone else" with all of the imperfection that this implies.

In several of the descriptions, there was a strong sense that the person "desired to remain disappointed." This is a phrase used by the psychoanalyst Charles Socarides[7] to describe how a person holds on to the disillusionment as a way to avoid facing the deeper implications of the experience. Focusing on a particular person one continues to regard as having unjustly failed to meet one's expectations may close off a number of difficult questions,

questions that most people who have been disillusioned would raise. These include, How did I come to hold such unrealistic expectations of the other person? Is my holding on to the disillusionment as an injury a way of not acknowledging and grieving for the loss of an illusion or set of illusions? Is there a sense in which I avoid taking responsibility for my own life by focusing on how the other violated my expectations, with the implication that I do not need to reflect on or question these expectations? One can also imagine, as the psychoanalyst Harold Searles suggests, that the angry or vindictive preoccupation with the failings of the other person may be a way of holding on to him or her.[8] Among other things, disillusionment tells us that the other is not "special" in the way that we thought he or she was, and, therefore, nor are we "special" in the way we wished we were. The dream that the special qualities of those we admire will "rub off" on us through close contact comes to an end.

Learning from the Literature on Disillusionment

In Chapter 1, I indicated that the experience of "seeing another as if for the first time" had been given little attention in the professional literature even though it is an important life event. In contrast, a number of psychologists and psychiatrists have written about disillusionment. As I discuss this literature I will be looking at the extent to which these authors help to deepen our understanding of the phenomenon of disillusionment as presented through the descriptions that I have discussed. It is not my intention to apply these theories to the stories in a detailed fashion or to "psychoanalyze" the individual stories. Rather, I want to bring the lived experience of disillusionment into dialogue with writers who have given considerable thought to this challenging human experience. However, this review of literature is by no means exhaustive.

These writers include a number of psychoanalysts, most notably Melanie Klein, Edith Jacobson, Heinz Kohut, and Socarides, who have written about disillusionment and idealization and all of whom are indebted to the work of Sigmund Freud. Whatever their differences, they agree that idealization and disillusionment are inescapable life issues. Disillusionment, they argue, may provide an impetus for us to move toward a greater sense of maturity and a stronger sense of self or it may give rise to self-protective cynicism.[9] The issue of disillusionment is also a key one for survivors of trauma, according to Ronnie Janoff-Bulman, and she is especially interested in how these survivors deal with what she describes as the "shattering of their world views."[10] There are also several phenomenological studies of disillusionment that provide important insights into this phenomenon.

Klein, Jacobson, and Kohut give particular attention to idealization, which, as we have seen, sets the stage for disillusionment. Unfortunately, it is difficult to summarize their views succinctly. To begin with, analysts make a number of assumptions (philosophical, cultural, and psychological), many of which are neither spelled out nor self-evident. Secondly, the language of psychoanalysis is to a large extent esoteric and technical. Freud is one of the few among analysts who had such a command of the language that he could write plainly and clearly when the occasion called for it. Moreover, analysts' conclusions are based on data or evidence that is obtained in a very particular context, namely, long-term intensive psychotherapy with patients. This kind of evidence is not directly comparable with what most of us learn about ourselves and other people through everyday observations and conversations. The argument can be made that their patients reveal to their analysts intimate aspects of their behaviors, thoughts, and feelings that even their best friends may not know about. A psychoanalyst could reasonably assert that the descriptions that I am working with do not provide accounts as complete as those that would have been uncovered in the course of psychoanalytic treatment. Some of the descriptions say little about the childhood origin of their idealizations, for instance.

On the other hand, psychoanalytic interpretations have traditionally tended toward reductionism. That is, they approach human behavior with a set of general assumptions, such as the existence of innate drives, that are applied to the lives of individual persons. Notwithstanding these differences, I hope to show that while phenomenology and psychoanalysis are distinct, they can be brought into fruitful dialogue with each other. Moreover, if one looks at some of the recent developments within psychoanalysis, it is apparent (as I will show in this chapter as well as in Chapter 4) that some psychoanalysts are increasingly concerned with interpreting human behavior in terms that are less abstract and more open-ended.

This dialogue becomes possible when one considers the specialized way in which analysts make sense out of what they learn about their patients. In everyday life we do not, as a matter of habit, interpret human behavior through the lens of psychoanalytic concepts and theories. If, for example, a friend tells us of his deep and unrestrained admiration for a woman he has only recently met, we would not quickly suspect that this admiration conceals attitudes, such as resentment, that are quite the opposite of admiration. Nor is it likely that we would speculate that this admiration might be related to childhood experiences of being frustrated or disappointed by his parents. In fact, I think it is fair to say that many people are skeptical about the validity of psychoanalytic theories. One objection is that these theories are based on work with disturbed people and therefore do not apply to the rest of us. However, I do not wish to quickly dismiss any consideration of the possibility that we, like patients in psychoanalysis, are neither as rational nor as self-aware as we would like to think that we are. My own view is that is that it is a mistake either to take psychoanalytic insights as established truth or to dismiss them out of hand.

How, then, am I approaching the challenge of discussing the psychoanalytic literature? To start with, I remind myself that psychoanalysts derive their insights from their work with and *experience of their patients* as well as their *experience of themselves* with their patients. What does this mean? Let us return to the example of the man who so admires (even idealizes) the woman he has recently met. We note that he is effusive in his expression of appreciation for her intelligence, wit, and sensitivity, her charming looks and graceful movement. She certainly sounds like someone who is admirable. But why does he have to go on and on about her? And why, we wonder, does he seem to be so unresponsive to the questions we ask about her and about his other feelings (beyond admiration) toward her? Where did she grow up, what are her interests, and what are her views on religion (a topic about which our friend has strong feelings)? Would he like to date this woman and is he sexually attracted to her? Our friend pushes these questions aside, perhaps even with an edge of irritation, as if they are beside the point. In turn, the more we hear about this apparently absolutely admirable woman (whom we have never met), the more we become annoyed with her as well as increasingly skeptical about our friend's unrelenting enthusiasm.

Faced with this kind of enthusiastic admiration, analysts are likely to start to think in terms of defense mechanisms such as reaction formation (perhaps this man unconsciously envies her success or resents her for being so "perfect" as to be out of reach) or denial (there is sexual attraction, but for some reason this man refuses to acknowledge it). They might also raise questions (in their own minds, at least) about our friend's relationship with his mother and, more specifically, about conflicting feelings he might have toward her. Is it possible that while he has positive feelings toward his mother at a conscious level, there are strongly negative feelings beneath these? Or perhaps his overall attitude toward her is one of deep disappointment, giving rise to a long-standing search for an admirable woman, with the twist that unconsciously he expects that these women will ultimately turn out to be as disappointing as his mother was.

Whether or not we are in agreement with the hunches or hypotheses that analysts might have about the meaning of our friend's behavior, there is the possibility of referring back to what is there both for us and for an analyst, in terms of our immediate experience. This is not to say that differences in interpretive perspective (everyday vs. psychoanalytic) do not affect our perception of events. However, there is something beyond interpretations to which observers with various perspectives can point.

Let me give another example, one taken from Klein's writings. In writing about what she and her colleagues have learned from observing infants, Klein quotes an analyst who uses the words "sleepy, satisfied, sucklings"[11] to describe infants feeding at their mothers' breasts. This is a description from an involved observer—the language is almost poetic, allowing us to visualize this interaction between mother and infant. In fact, Klein speaks of how one has to be in sympathy with the infant, with one's own unconscious in close touch with the infant's unconscious, to arrive at an understanding of his or her situation. In ordinary language, one could say that Klein advocates using one's intuition. But beyond description, there are interpretations and hypotheses, and generally Klein is quite clear in distinguishing between the behavior patterns she observes and what she believes they mean in terms of her theory of human development. In regard to the description of feeding given earlier, Klein provides an interpretation of how these infants have dealt with what she assumes to have been their initial anxiety toward their mother's breasts. Reading Klein, one notes that there is something that she (and her colleagues) observes, and that we too would observe, and that there are also interpretations that Klein comes up with that we might or might not agree with, or that, at the very least, are not self-evident.

Along this line, Constance Fischer, a phenomenological psychologist who has written extensively on psychological assessment, makes the distinction between primary data, that is, what we experience and observe directly, and secondary data, that is, the conclusions we draw on the basis of this data. She makes two points about this distinction. The first is that we often forget that there is a difference between primary and secondary data. Thus, a psychologist does not "observe" a person's IQ (intelligence quotient), but does observe the particular way and the success with which the person tackles the tasks that make up an intelligence test. Her second point, and the one that is less obvious, is that secondary data are not more "real" than primary data. In fact, the opposite is the case. That is, IQ is not a fundamental capacity or trait in the person that causes him or her to be smart. Rather, IQ is a score that is attributed to the person on the basis of how well he or she handles different kinds of structured problems and questions on a test.[12] The same point applies to psychoanalytic concepts and interpretations—they are not more basic or real than the life of the person whose behavior they intend to explain. Their value and usefulness depend on how well they help us to understand and deal with people. And the fact that one has one set of useful interpretations does not exclude the possibility (or even the likelihood) that another set can also be useful, fitting, and appropriate. It is with this basic understanding of the relationship between experience and interpretations that I approach psychoanalytic theory as well as other theories.

In discussing the relationship between theory and experience, I am being pragmatic or practical. That is, I am concerned with the extent to which theory helps us to understand human experience, to see it in a new way, or to give attention to aspects of it that might easily be overlooked. By "understanding," I do not mean just an intellectual appreciation of what is going on, but a deeper, more felt sense that something falls into place so that we now *grasp* more of what is going on at a personal level. We can speak of a particular theory as having heuristic (derived from the Greek word for "discovery") value, which is to say that it gets you somewhere. The social psychologist Kurt Lewin wrote many years ago that there is nothing as practical as a good theory. But there are two risks to keep in mind about theories in terms of their practicality. The first is that we can readily be seduced into believing that good theories give us the truth. The other danger is that in our eagerness to be pragmatic and helpful, we rely on theories to tell us what to do and fail to base our interventions on a solid understanding of the individual situation.

As I proceed with connecting theories back to experience and looking for their heuristic value, I will try to avoid getting caught up in issues that are technical and of interest primarily to psychoanalysts. Obviously, there are limitations in this approach. Perhaps, for example, I might distort the meaning of someone's concepts by considering them outside of the context of this person's overall work. However, my foremost agenda is to further an understanding of the experience of disillusionment, not to consider specific psychoanalytic theories on their own terms.[13]

I start with Klein (1882–1960) because of her emphasis on idealization. She has been a highly influential analyst, especially in England and in South Africa. Early in her career she viewed herself as an orthodox follower of Freud but, by the time she was in her forties, it was evident that she was an original thinker whose position was at odds with Freud's in several significant areas.[14]

Klein believed, as did Freud, that early childhood experiences are crucial in the development of personality, that unconscious factors play a large role in human behavior, and that interpretation is a powerful technique in analysis. She also agreed with Freud that there are two powerful innate drives in human beings, eros and thanatos, that is, the life instinct and the death instinct. Both Freud and Klein, along with most psychoanalytic thinkers, can justly be described as *intrapsychic* theorists. This means that they conceive of the person as an individual and as the primary unit of analysis. It is not as if they thought environmental factors were unimportant, although neither gave much attention to influences such as peer relations, experiences at school, or conditions at work. To a large extent, however, they conceived of the environment as acting on a personality structure that somehow existed apart from the environment. Thus, when reading Klein, one is struck by the detailed analyses of children and their traits, whereas consideration of current relationships with parents and the child's social environment take a back seat.

Unlike Freud, Klein treated young children and very seriously disturbed and sometimes psychotic adults psychoanalytically, and she made systematic observations of infant behavior. Freud's view was that children do not have sufficient ego development to process psychoanalytic interpretations. However, he made this argument on theoretical grounds. Klein, on the other hand, argued from her own and her colleagues' experience that interpretations could be used effectively with children. She wrote that if interpretations were expressed in plain language, they often produced noticeable improvement in the child who was greatly relieved at being understood by a caring and reliable adult.[15]

To fully appreciate Klein's view of idealization, we first have to take into account how incredibly vulnerable and dependent she believed infants to be. At first sight, this is hardly a novel idea. Anyone who has observed infants even casually recognizes that they are radically dependent on their mother or caretaker for their physical and psychological survival. The distress of infants who are hungry or who are separated from their mothers, even for a short time, is palpable. But in Klein's mind the infants' situation is even more difficult than is apparent from everyday observation. As we have seen, she agreed with Freud that we are born with innate drives, one of which is the death instinct. Freud conceives of the death instinct as directed toward the self, that is, as self-destructive energy. Klein discusses at length how, from the very beginning of life, we are saddled with the daunting task of managing this destructive energy, drawing upon the life instinct to contain it.

Although there are problems with the assumption that humans have a death instinct—it is a controversial concept even among psychoanalysts—it would be a mistake to dismiss Klein's theories out of hand. One of Klein's strengths is that she (along with Freud) confronted the reality of human destructiveness very directly; for her, hate was as central to life as love. Klein and Freud lived through two world wars that brought unimaginable destruction to European countries, and so, to them, the idea of a death instinct was all too plausible. And tellingly, Chris Hedges, a contemporary observer of human destructiveness, concludes his book about his experiences as a war correspondent with a chapter entitled "Eros and Thanatos."[16] This is not to say that he endorses a Freudian view of instinctual drives, but that he clearly regards them as meaningful metaphors.

Obviously, no one could reasonably accuse Klein of having a sentimental view of human nature or of the relationship between infant and mother. Yet it would it be unfair to portray her as having a one-dimensional and simply negative view of human beings. As Meira Likierman points out in her study of Klein's life and work, Klein's last publication dealt with the human search for companionship, and here she "returns to a more compassionate, poignant appraisal of the human psyche."[17]

To understand adult behavior, Klein asserts, one must first understand the relationship of the infant with its mother and, specifically, its relationship to the mother's breast. The infant requires the availability of the mother's breast not only for its physical well-being but also for its psychological security and its sense of being loved and being worthwhile. In her emphasis on the infant's psychological well-being, Klein is again departing from Freud. In her view, the infant seeks a personal connection to others from the very beginning of life, whereas for Freud the connection is primarily instinctual at this early stage. According to Klein, children have an innate capacity for love, along with the capacity for greed and envy. [18]

Infancy is a state of great vulnerability since the infant is dependent on its mother while having no direct control over her behavior. However, analysts argue that the infant can exercise some degree of control over its own experiencing and mental processes by way of *defense mechanisms*. Defense mechanisms operate outside of consciousness and "transform" inner and outer reality. That is, we can deny aspects of the world that threaten us and we can repress or otherwise evade our own desires and memories. Klein postulated that infants, who are terrified of the self-destructive instinctual forces within themselves, manage these by *projecting* them onto outside targets, most notably the mother's breast. She believes that it is less anxiety-provoking to confront external than internal threats since one cannot escape from, or retaliate against, dangers that exist within the self. This line of interpretation may seem far-fetched and it may seem especially implausible, given that all of this is taking place in the first year of life. Even Freud, who was no stranger to attributing complex psychological reactions and relationships to children, thought that Klein was overestimating the psychological maturity of infants. My main concern here, though, is not so much with the adequacy or plausibility of Klein's overall theory of development as with the potential value of her insights into idealization.

There is one more basic issue we have to look at before I can spell out her position on idealization. How can the infant live with an "angry" breast, or, in psychoanalytic terms, a "bad breast"? If the infant—and not just the disturbed infant—experiences its own aggression as coming from the mother's breast (adding to whatever irritation or anger the mother may "actually" at times feel toward him or her), the same breast that the infant absolutely depends on, then, surely, life would be intolerable. Such a grim view of infancy is obviously highly one-dimensional. Anyone who has observed infants being breast-fed knows that breast-feeding is often pleasurable for both mother and child. Klein would agree, but she believes that there is another defense mechanism at work, practically from the beginning of life, namely, *splitting,* that, in a manner of speaking, saves the day. Splitting means the infant divides up the mother, so to speak, into the good breast and the bad breast and also separates its own aggression from its loving side.[19] There is a categorical separation that is only overcome gradually and partially as part of the child's (and adult's) maturation. The concept of splitting has been accepted by many within the psychoanalytic community, including some who reject the concept of the death instinct, as a powerful way of pointing to how we deal with a variety of situations, including children's relationships with their parents, patients' relationships with their therapists, and nations' dealings with one another. Many of us remember how President Bush continually presented the war on terrorism in terms of a war of good versus evil subsequent to the attack on the United States on September 11, 2001.

With this background, we can outline Klein's interpretation of idealization, which, in infancy, takes the form of idealization of the good breast. For her, there are three motives or purposes at work:[20] idealizing the good breast is a way of protecting oneself against the bad breast. That is, the more wonderful, powerful, and generous the good breast (and, more broadly, the mothering person) is thought to be, the more this idealized other is a protection against anything dangerous, threatening, or depriving. The threatening forces include one's own unconscious aggressive and ambivalent feelings toward the mother. It is, after all, largely the projection of one's own destructiveness onto the other that makes the other (the bad breast) seem so dangerous. Given the degree of dependency and vulnerability of the infant, it is fair to say that idealization arises out of fear and desperation. This suggests that the degree of idealization is proportionate to the degree of desperation.

But fear is not the only motivating force. Idealization also serves desire, since infants hunger for unlimited gratification. And not just infants! Later in life, the idealization of a lover as infinitely understanding and attentive may be an expression of one's desire to be endlessly admired and catered to. Finally, idealization may involve the projection of one's own goodness or loving capacity onto another. It is noteworthy that Freud did not attribute such a capacity to the infant. Although he and Klein readily affirm the darker side of humankind, Klein also affirms more positive human inclinations. The problematic consequence of this projection of one's good qualities onto another, she notes, is that it conceals how they are part of oneself. To put it differently, our excessive admiration of the other may be accompanied by a diminished valuing of ourselves.

This splitting and idealization characterize the early phase of infancy. Klein calls this the "paranoid-schizoid position." In their discussion of this position, Stephen Mitchell and Margaret Black explain succinctly why Klein uses this terminology.

> Paranoid refers to the central persecutory anxiety, the fear of invasive malevolence, coming from outside [. . .] Schizoid refers to the central defense: splitting, the vigilant separation of the loving and loved breast from the hating and bad breast [. . .] The bifurcated world of good and bad was not a developmental phase to be traversed. It was a fundamental form for patterning experience and a strategy for locating oneself, or, more accurately, different versions of oneself, in relation to various types of others.[21]

Using the language of psychopathology to describe human development in general is problematic; perhaps Klein did not see this as an issue because her whole life work as an analyst was with seriously disturbed persons. Putting this objection aside, it is important to note that this position, or form for patterning experience, is one she believes continues into adult life.

Klein's second stage, the depressive position, involves coming to a more realistic perspective regarding oneself and others. The infant—still less than a year old!—comes to recognize that there is not a good and a bad mother or a good and a bad breast. Its illusion of an idealized mother/breast collapses, and this loss brings about grief and depression. Yet this is by no means just a negative state of affairs because under favorable circumstances the child recognizes and regrets its own destructiveness and wants to repair any damage that it has caused to its relationship with the mother.[22] This position (as is true of the paranoid-schizoid position) is not one that any one of us leaves behind once and for all. However, part of what does enable the infant to move ahead is that "his growing capacity to perceive and understand the things around him increases his confidence in his own ability to deal with and even to control them, as well as his trust in the external world."[23] So, for Klein, the ongoing exploration of the world is what allows the child to leave behind, however incompletely, the distorted views of the previous positions.

Jacobson (1897–1978) has also made an important contribution to our understanding of idealization and disillusionment. She accepted Freud's (and Klein's) assumption about the existence of the instinctual drives of the life and death instincts but arrived at a different view of their nature. For Jacobson the child's interaction with parents is no less important than innate predispositions. She regarded the instinctual drives as plastic rather than fixed in nature—the development of the drives as partly shaped by the child's relationship with others—and their evolution as codetermined by the child's relationship with others.[24] Moreover, she saw these supposedly opposing forces of destructiveness and love as having the potential to interact in constructive ways. Thus, while the desire for love (and union) with another might push a person toward getting overly involved in a romantic relationship, the aggressive tendency can counteract this tendency by pulling away or separating from the other person. In this context, aggression preserves the person's sense of identity. Overall, Jacobson regards this instinct as a potential asset rather than as a fixed liability that at best might be contained.

Jacobson is classified as an *ego psychologist*.[25] For Freud, the ego was the part of the personality that was involved in dealing with reality (e.g., thinking, judging, remembering) and with managing the tension between desires, many of them unconscious (the id), and the person's judgments, about right and wrong (the superego). Freud introduced the concept of the ego relatively late in his career and acknowledged that it needed further development and study. The ego psychologists took up this task, bringing with it an overall shift in emphasis. Rather than focusing primarily on unconscious conflicts, these analysts were very interested in what Jacobson called *psychic structure* and how this structure develops as the child grows up. Simply expressed, psychic structure refers to the people's capacity to manage their own emotional life, relate to others, take obstacles and opportunities into account, and respond to disappointments and conflicts. An increasingly clear but not rigid sense of identity is one sign of healthy development, or "ego strength." Whether or not there is healthy development depends on relationship with parents and other environmental as well as constitutional factors.

It is within the context of the development of psychic structure that Jacobson discusses idealization. She believes that there are three motives for the idealization of parents in early childhood.[26] Along with Freud and Klein, she viewed the child as frightened by its own ambivalent, and even sexual, feelings toward parents. In the classical Freudian perspective the boy desires his mother and the girl her father. Idealizing parents, in a sense, puts them out of harm's way, or, more accurately, puts the child's relationship with the parents out of harm's way. By seeing the parents as wonderful and powerful, the child pushes its ambivalent feelings out of the way and asserts the opposite, namely, that they are just strong, positive feelings. Moreover, according to Jacobson, parents live "inside" the child, in addition to existing as separate beings in his or her world. Just as there is an intimate physical contact with the mother during breast feeding, so there continues to be an intimate psychological connectedness with the mother (and the father) throughout childhood and, in more subtle ways, throughout the rest of one's life. So, to have wonderful parents is to bask in the glow of the parent's wonderfulness. This means that in addition to strengthening the child's sense of security in relationship to the parents, idealization broadens the child's image (I am the son or daughter of so and so) and also raises his or her self-esteem (my parents who really love me are wonderful and powerful people, and they will tell me so even when someone else doesn't like me).[27]

There are several other points that Jacobson raises that I want to comment on here. These have more to do with the aftermath of disillusionment than with idealization per se. She believes that disillusionment contributes to maturation, providing it does not undermine the child's confidence in caring adults, especially its parents. If these disappointments are too severe, an attitude of cynicism is a likely outcome. Otherwise, they provide the child with occasions for gradually abandoning a world of magic and developing ideals that are viable guides for everyday behavior. The notion of leaving behind a world of magic also comes up in the work of Bas Levering, an educational researcher. Proceeding from an analysis of the concepts of disappointment and disillusionment in everyday speech, he concludes that "disillusionment shows that the world is quite different from what had been thought or believed,"[28] in the sense that there is a loss of enchantment or a spell that is broken.

So disillusionment, under the best of circumstances, may lead from magical illusion, of thinking of significant others and oneself as "special," to living with values and ideals that are more realistic. However, one's ideals, principles, and self-expectations (the part of the superego or conscience that Freud calls the ego-ideal) are based on idealized images of self and others. This is inescapably how ideals develop, from a psychoanalytic point of view. However, there is a potential problem in this process. Jacobson quotes Freud to the effect that there is the possibility that the child does not give up his idealized image of himself (and/or of others) but transfers these images to his ego-ideal. In plain English, there is a risk that while some people superficially give up regarding themselves as being special—what

Freud calls the narcissism, or self-love, of childhood—in reality this special-ness is just transferred from their image of themselves to the ideals that they uphold.[29] In other words, holding oneself to high ideals—presumably a good thing—may, upon closer inspection, turn out to be another version of holding oneself up to be special—another way of being self-preoccupied, except, this time, in a form that is more easily hidden from both self and others, and that is more easily defended. Some people who make a point of "standing on principle" may fit this profile.

In a 1977 article entitled "On Disillusionment: The Desire to Remain Disappointed," Socarides[30] picks up on some of these themes regarding the aftermath of disillusionment. (Regrettably, Socarides and Kohut use the words disillusionment and disappointment almost interchangeably. This is unfortunate, because the distinction between the two concepts is clear in everyday usage: I am disappointed if a good friend is late for a lunch meeting, but I am disillusioned if he steals my assets and runs off with my beloved!) He identifies the core of disillusionment as a violation of a strongly held expectation, with the "consequent loss of ability to find value and inter-est in things as they actually are."[31] Ideally, with time, the outcome is a more realistic view of the world, one that does not preclude mature hope and trust. But there is also, as Jacobson points out, the possibility of chronic and embit-tered cynicism as a protective stance to ward off further disappointment. Following the philosopher Gabriel Marcel's analysis of hope, as I discussed it in Chapter 1, one could see this transition as a transition from optimism (a kind of naive assurance about people's positive qualities) to pessimism (an equally narrow conviction that other people are not trustworthy). A more mature stance would include a recognition of the varied qualities of human beings, including oneself, and an openness to a variety of outcomes in a relationship. From Socarides' perspective, as well as from the perspective of most analysts, negative outcomes are thought to be largely due to a lack of a secure bond with others in childhood.

As we have seen, there is considerable overlap among the theories of Klein, Jacobson, and Socarides. With Kohut (1913–1981), there is more than a shift in emphasis. Although he started out as an orthodox analyst, he eventually broke away from Freud's emphasis on instinctual drives. Instead, he came to believe that the most basic psychological issue was the develop-ment, protection, and enhancement of a person's sense of self. Further, this development must be understood as taking place in the context of the per-son's fundamental relatedness to, and dependence on, others. His definition of psychological or self structure, as meaning "nothing more than having a formerly external function permanently in your possession,"[32] indicates more specifically how the self is dynamic and depends on relationships. According to Kohut, this reliance does not end with childhood, but continues on, although in a more muted form, into adulthood.[33] In his work with patients, Kohut gave great credence to the power of the therapeutic relationship and to the value of interpretation.

But by interpretation he meant something different from most analysts. It was not a matter of analyzing, for example, the unconscious conflicts that gave rise to the patient's problematic behavior. Rather, in a more phenomenological vein, Kohut tried to understand and express what he thought were the patient's personal meanings at a given point in time. So, essentially, analysts ought to orient themselves to their patients' point of view; in a word, they ought to be empathic. This concern with being faithful to the point of view of patients also extended to Kohut's theorizing because he viewed analysis as being scientific insofar as it accurately grasped what was going on for human subjects. He rejected the assumption of a death instinct because it did not fit with his clinical experience. Aggression, he concluded, was secondary to disturbances in the sense of self. It was a signal of disturbance, not the cause of it. Kohut's perspective was close to phenomenology in that the concepts he developed were ones that he believed were critical to understanding and successfully listening to what was going on for patients.

One of his key concepts was *selfobject,* a term that refers back to the way in which we are reliant on others, as indicated earlier in this chapter. For example, a nine-year-old girl returns home after a soccer game that her team lost. She is feeling demoralized and looks to her father for encouragement. From her perspective she receives such support when he says that it is too bad that her team lost, but that he knows that she did her best and that is all that anyone can expect. This conversation helps her to feel better about herself and her team's defeat. Later in life, she likely will be able to boost her own morale, but at nine she looks to others for help. Kohut would say that the girl's experience of her father played *a selfobject* function insofar as her interaction with him helped to bolster her sense of herself. In other words, there are selfobject dimensions to any interaction where the other person (the "object") contributes to our sense of psychological continuity, well-being, and maturation (all of which are ways of talking about the "self"). Kohut believed that this process takes place unconsciously earlier in life but is available for reflection by adolescence or adulthood.[34] He also made a number of distinctions in terms of how various interactions with others affect the development of the self. Here I will only mention two: mirroring and idealizing selfobject experiences.

Mirroring is illustrated by the example of the girl's interaction with her father. The significant other is experienced as affirming one's value, energy, and specialness, as in the case of the mother who applauds her young son when he demonstrates his newly acquired skill in reading. What is significant here, Kohut recognizes, is not what the mother does per se, but the child's recognition and "taking in" of her valuing and appreciation of him. Such interactions allow the child to grow up feeling important and loved and to withstand criticism or disappointments later in life. The idealizing selfobject is more relevant to the issue at hand. The child ordinarily idealizes its parents, as Jacobson and Klein have also emphasized. For Kohut, the process of idealizing parents and then eventually being disappointed by them can contribute to the process of the development of the self. This idealizing provides the occasion for the child to incorporate and be inspired by values, goals, and ambitions that will help to make his or her life meaningful. First, the parents are looked up to because of their perceived special values and powers. Then, assuming that the process of being disappointed does not come too abruptly, the child internalizes some of these perceived values and goals and tests them out over time so they become increasingly realistic and viable. Problems arise if parents do not allow their children to idealize them or give their children little reason for looking up to them.

Kohut's view of idealization is distinctive. He regards it primarily as a creative and necessary way of enhancing one's own "psychic structure" and only secondarily as a defensive maneuver (e.g., a way of concealing one's aggression toward the parent). In psychotherapy then, the idealization of the therapist is also regarded as potentially constructive. Unlike many of his colleagues, Kohut believes that the therapist should allow this idealization rather than attempting to undermine it through interpretations. The idealizing transference, as Kohut calls it, provides a way for patients to develop some of their own ideals and make up, to some extent, for lack of opportunities to idealize parents. Once having been permitted to idealize their therapists, these patients will, over time, come to see their therapists' fallibility as well.[35] Hopefully, this process takes the form of tolerable disappointments that can be managed within the therapeutic relationship.

The theme of dealing with loss and trauma has been taken up by Peter Homans, a sociologist who has been influenced by Kohut. Homans discusses the themes of disillusionment, mourning, and creativity in *The Ability to Mourn,* a book on the origin of psychoanalysis.[36] His thesis, greatly simplified, is that the creativity of Freud and other innovators within the psychoanalytic movement was a response to personal and collective loss. At the personal level, grief arose from events such as Jung's separation from Freud; at a collective Western European social and cultural level, there was the loss, dating back to the fourteenth century, of a sense of communal wholeness and a coherent world view.

In her book on trauma, Ronnie Janoff-Bulman also emphasizes the loss of a coherent world view, or, as she puts it, the shattering of world views. In words that echo Socarides, she writes about the aftermath of trauma: "Victims' inner worlds are shattered, and they see their prior assumptions for what they are—illusions. In the end, the adjustment of survivors rests largely on whether they experience profound disillusionment and despair or, ultimately, minimal disillusionment and hope."[37] By trauma, Janoff-Bulman and her colleagues mean that caused by events such as rape, life-threatening diseases, death of a loved one, and debilitating accidents.[38] While disillusionment is not ordinarily equivalent to trauma, the person's fundamental assumptions are brought into question in either case and, thus, there is significant loss involved. In a recent publication, coauthored with Michael Berg, Janoff-Bulman discusses the existential gains that may come as the consequence of loss. Trauma destroys comfortable illusions, such as the belief that accidents and crime happen to other people, and thus brings home to people how vulnerable they are and how precarious life is. The realization of life's precariousness may also lead survivors to a recognition of its preciousness and thereby to a greater appreciation of its value. Accordingly, the authors conclude, "It is knowing the possibility of loss that promotes the gains of victimization, and that of disillusionment that creates a newfound commitment to living more fully."[39] Several of the analysts I have discussed make similar assertions although they state their position less plainly.

There is also an emphasis on outcome or adjustment in three phenomenological studies (in the form of dissertations) on disillusionment in everyday life. Vernon Holtz examined the adult experience of disillusionment in the context of religion, marriage, or career, while John Neubert looked at disillusionment in midlife, and Christian Carson Daniels focused on young adults of "Generation X."[40] Holtz found that disillusionment in the context of career was less disturbing than in the context of marriage or religion. Several of the nine people he interviewed took ten to fifteen years to arrive at some degree of peace with their loss of illusion. Only one of Neubert's six interviewees was able to let go of his previous idealized vision and find new meaning in his life. This man went through a process of deep despair, letting go of his previous idealization, and then allowed himself to open up to a woman who loved him.

Holtz and Neubert essentially agree that "to be disillusioned means the shattering loss of unquestioned, rigidly lived beliefs present within a crucial-identity-oriented relationship."[41] The rigid belief involved an idealization, whether focused on a relationship, a religious belief, or a way of living. For example, one subject had believed if she lived in a certain way, God would respond to her prayers in the manner that she desired; another took it for granted that her husband would automatically be faithful to her because she took care of him and was a loyal wife. These beliefs were taken for granted so much that the subjects did not realize that they were beliefs; they were taken to be the way things were. Along the same line, Carson Daniels found that the resolution of disillusionment involves the recognition that all beliefs are beliefs rather than truths. Because of this assumption, these two researchers also described the beliefs as involving inauthentic hoping. By this they meant that these anticipations of the future ignored the essential ambiguity and uncertainty of life. Neubert in particular emphasized that their anticipations negated the separateness and autonomy of the other person involved in the relationship. In other words, "the significant other/object existed for the subjects as extensions of their respective vision,"[42] or, to use Kohut's terminology, in the relationship the other functioned primarily as a selfobject. Thus, there was a tacit assumption that the way one insisted things turn out is what would actually happen. For this reason, the word "hoping" is misleading because hope is by no means equivalent to an expectation. Hope, as Marcel, suggests, is an attitude of openness to life's possibilities.

Finally, a "crucial-identity-oriented relationship" means that subjects believed that their value as persons and the meaningfulness of their lives were fundamentally dependent on things turning out according to their preconceived idealized and illusory visions. Unless they could find a way to let go of these visions, they would live out their lives in an attitude of chronic disappointment and cynicism. Whether letting go occurred depended on what each person experienced (or allowed himself or herself to experience) after the disillusionment. As I have mentioned, only one of the six people Neubert interviewed was able to move beyond his original vision, and this happened as he became involved in a loving relationship.

Carson Daniels did not find that the basic of disillusionment was different among members of Generation X as opposed to other generations. However, her findings in terms of outcome were more encouraging than those of either Neubert or Holtz. That is, she found more cases of resolution among her subjects (as did I in my study) than was the case for them.

Dialogue between Stories and Literature

What do we find if we bring these thinkers into dialogue with the stories and the phenomenological analysis I presented at the beginning of this chapter? Perhaps the best way to begin is to consider the context and motives involved in idealization. As we have seen, psychoanalysts believe that idealization first arises in infancy/early childhood when children are highly dependent on adults and that it serves defensive (or protective) as well as self-development functions. For Kohut, especially, but also for Jacobson, idealization is an essential process that contributes to the growth of psychic or self structure. However, the descriptions that I have collected do not, and could not, address early childhood origin, even though some refer to childhood experiences. How could an adult recall experiences from infancy or how could he or she describe processes that presumably were largely unconscious? Of course, these questions are not so easily answered by psychoanalysts either. The developmental theories of Freud, Kohut, and most other psychoanalysts are primarily based on their interpretations of patients' recollections of their childhood and, thus, are removed by several degrees from the actual experiences of children. Klein observed and worked with young children but, as she herself acknowledges, her interpretations of their world are rendered in adult language and fashioned by way of psycho-analytic concepts. In this regard Silvano Arieti and Jules Bemporad, analysts with a phenomenological bent, have commented that developmental theo-ries are "adultomorphic" in the sense that they say more about the mind of the adult than the world of the child.[43] However, these difficulties do not preclude a meaningful dialogue between the insights of psychoanalytic theo-rists and phenomenological analyses based on descriptions of everyday life experience. Psychoanalysts believe that the events of childhood leave endur-ing impressions and that the dynamics of adulthood recall those of earlier life. As we have seen, Klein, for example, asserts that the paranoid-schizoid and the depressive positions persist, in some form, into adulthood. With this in mind, let us look at these stories and consider to what extent psy-choanalytic and other theories can enrich our understanding of disillusionment.

Is it fair to say, on the basis of the descriptive data, that idealization arises in the context of dependence and vulnerability, as all of the psychoanalytic theories insist? The answer is clearly yes. As I stated at the beginning of this chapter, the identity (or sense of self) of the person who idealizes and his or her hopes for the future are closely intertwined with the person who subsequently lets him or her down. The stories describe idealization of fathers, psychotherapists, older friends, lovers, and mentors, all of them people who play critical roles in the narrators' lives. Some of the stories illustrate this critical role quite dramatically. Jasmine's father saved her and others from drowning. Joan, who later became disillusioned with her therapist, started therapy at a point in her life when her marriage was coming apart, when she felt estranged from herself, and she had serious doubts about her own ability to function as a mother. Margaret believed that her stepfather, David, gave her the opportunity to belong "to a real nuclear family" and to feel loved at a time in her life when, as an adolescent, fitting in was especially important. For Ellen, who had idealized her father, his support was essential since she experienced her mother as denigrating her talents and her value as a person. She herself affirmed that "my father has always been very tied up in my view of myself," reminding us of Holtz and Neubert's discussion of a "crucial-identity-oriented relationship."

In many of the stories, one is struck with how the idealization points to who the person yearns to become. That is, the relationship seems to be focused not so much on what has already happened as what might happen in the future. Rosemary's admiration for the Episcopal priest was based on what this woman could do, that is function successfully as priest, a possibility that was closed to her as a Catholic. The idealization was not based on any support Rosemary received from this woman, although she initially thought the priest was very nice and personable. This story brings to mind Klein's concern that when we attribute strength and goodness to others (qualities that we have without recognizing it consciously), this may well diminish the self. However, the determining factor is how the relationship develops over time. As Jacobson and Kohut suggest, idealization of someone powerful (especially parents) can serve to enhance one's sense of self. But such enhancement presupposes an ongoing and positive relationship and a gradual transition to standing on one's own feet. Rosemary did not have an enduring or positive relationship with the priest whom she idealized. At the same time, by dealing with her disillusionment, she did come, however painfully, to a greater appreciation of the value of her own judgment. Her conclusion that "priests are like everyone else" implies that priests are not beyond or above her. She also spoke of how she would be better prepared to evaluate any future applicant for the same position, which meant that she now saw herself more explicitly as someone who evaluated others rather than someone who was primarily seeking approval.

Jeannine wished that her sister Doris would send her daughter to regular school so that the girl could break away from a family pattern of isolation. If Doris were to do so the precarious sense of family that Jeannine had begun to experience might continue. In Benjamin's relationship with the woman he admired, there seemed to be a strong emphasis on what was yet to come; he imagined they might have a special future together but it was not clear that he saw this future as grounded in their current relationship. In contrast, Richard's admiration for his older cousin was based both on past and future; his cousin was a model for what growing up meant, and had been so since Richard was six years old. The cousin showed Richard how to deal with things in the present and provided him an image of what he wanted to become. On the basis of this overview, it is fair to say that idealization arises in a situation of vulnerability and dependence, and that, generally speaking, the younger the person the greater the dependence. Idealization also seems to be intertwined with strong hopes or wishes for own one's future, with the implication that one's current situation is neither satisfactory nor secure.

Klein argues that idealization protects the person from the threat of anger or aggression, especially one's own aggression as projected upon the other. As part of this defensive process, there is a splitting of the object into the "good" and the "bad." As I have indicated, it is not possible to directly confirm or disconfirm her theories about infancy on the basis of descriptions written by adults, but it is possible to examine whether they apply to the stories that people tell about their present circumstances. Ellen's story is the one that most obviously brings "splitting" to mind since she regarded her father as the good parent and her mother as the bad one. In Klein's view, splitting can take the form of equating one person with goodness and another with badness. But what Ellen's story also highlights, and what is easily overlooked, given Klein's emphasis on the individual, is that "splitting" is not just a matter of what individuals do within their own mind, but that it arises in an interpersonal context. As Ellen came to realize as a young adult, her father let her mother do the disciplining, and her mother allowed the father to be the parent who was supportive, and so "splitting" had been a family affair.

The long and painful process Ellen went through as she became disillusioned with her father resembles Klein's description of the depressive position, where splitting is overcome. That is, Ellen started to see her father as fallible and weak in some ways. At the same time, she also began to experience her mother as more human, as being frail and not just malicious. She started to blame her father for some of the problems in the family and she pulled back from him for some time. Eventually, she said, "I see my father again, as a totally new person; and get to know him again. And then I started to see the good parts again." Some of the other stories follow a similar pattern. Consider Joan's relationship with her psychotherapist. Over an extended period of time she came to think of this former therapist as a woman with her own problems and issues. But this process was made possible through long-term psychotherapy, with another therapist, and took place as she made other changes in her life. In Jasmine's account her good father turned into the bad father who betrayed her. For her, as well, the idea of "splitting" seems relevant. It was her father who was the good parent, while her mother was the one who was to blame for arguments between the parents. When the parents separated, Jasmine also blamed her mother, and this did not change until she overheard her mother telling a friend that her husband had been involved with another woman for a number of years.

It was only with the help of a priest who counseled her that she found a way to recover her relationship to her father as a good man and forgive him for how he had hurt her and her family. Klein states that in the depressive position, there is realization of, and regret about, one's destructiveness toward the other. In these two accounts there is regret and self-blame for having been too gullible and, more implicitly, regret at having been so critical of the one parent while idealizing the other.

It would not be unreasonable to suggest that in idealizing her father, Ellen protected herself. Faced with someone who was a formidable threat, any of us would surely be eager to find an advocate or protector who was truly wonderful and strong. This may have some analogy to the tendency of the public at large to idealize a national leader in time of war or threat to the nation. President George W. Bush's rapid increase in popularity after the 9/11 terrorist attack appears to be more than an expression of appreciation for how he responded as commander in chief in the months immediately after the attack. In other words, it did not seem to reflect a balanced assessment of his overall effectiveness and accomplishments. If we are in danger, we need a strong leader, and thus, psychologically, we can ill-afford to look too closely at his or her limitations. So, even if one rejects Klein's assumption of a death instinct, it is possible to see how idealization might hide or deflect aggression or criticism toward the person one idealizes.

In a similar vein, Jacobson has suggested that our idealization of our parents shields us from our ambivalent feelings toward them. For the reasons discussed earlier—how can we remember early childhood experiences and how can we describe what presumably is unconscious—this hypothesis is difficult to evaluate precisely. Nevertheless, it is an interpretation that makes sense intuitively. By definition, idealization means that we regard someone as ideal or perfect; then, how could one have ambivalent feelings about someone who is that wonderful? But then again, how could one not, especially if the person one idealizes is someone on whom we depend and is someone we have a lot of contact with, such as a parent? To idealize someone is to construct our relationship with him or her along specific lines: the other is powerful, helpful, someone to look up to. In this type of relationship we are the one looking up, the one being helped, and the one who is not powerful but lacking in some way. And surely the people we idealize disappoint us long before we become disillusioned with them. Yet, characteristically, these disappointments are swept aside as having no real significance. In contrast, when there is an absence of idealization the mixed feelings toward the other on whom one depends may be more in evidence, as we have seen in Jeannine's relationship with her sister Doris. Generally, though, the person who does idealize someone will minimize or refuse to acknowledge feelings other than those that are positive. This is what Holtz and Neubert call rigidity. It is this very refusal, along with the person's one-dimensionally positive description of the person they admire, that gives the listener occasion to pause and wonder about what else is going on in the relationship.

However, clearly, Jacobson and Kohut see idealization as doing more than protecting the child from difficult and ambivalent feelings. Whether one uses the language of identity, self structure, or the language of one of the women I interviewed—"My father has always been very tied up in my view of myself"—it is evident that idealization is a vehicle for enhancing one's self-experience. Or, to use Kohut's language, the idealized other functions as a selfobject. Ellen describes how her father both affirms her value as a person (acts as a mirroring selfobject) and helps to provide her with a vision of herself as successful in her future profession (idealizing selfobject) just as he, the father who admires her, has been very successful in his profession. For Rosemary, the Episcopal priest she admired embodied her hopes and aspirations for herself, and yet this priest provided no positive response to Rosemary as a person. One wonders to what extent any of us can successfully incorporate ideals from someone who does not in some way respond in an affirming way to us. This is a theme that all of the psychoanalysts I have discussed address. They believe, reasonably enough, that the bond between infant and parents is critical not only in allowing for idealization but also for making it possible for the child to move through the process of disillusionment and disappointment later in life. They recognize, of course, that children often idealize parents even when they are abusive (and even abusive parents sporadically treat their children with concern), but are keenly aware that, in these cases, the long-term prospects for the children's coping with their disillusionment are much less favorable.

Margaret felt respected by David, her stepfather, and relished the way in which his care and availability gave her the sense of being a normal teenager with a dad. But none of this prepared her for David's breakup with her mother and his disappearance from her life. No doubt the preexisting problems in her family had already left her feeling very vulnerable. This experience of having one's most heartfelt hopes raised and then dashed would have been very difficult for anyone to manage, especially during adolescence. As a patient in psychotherapy, Joan was also vulnerable. She experienced her therapist as warm and caring and felt accepted and validated by her. Again, in Kohut's terms, she experienced her therapist as a mirroring selfobject. As her therapist's behavior changed, and especially as her therapist insisted that she see her husband instead of her, and then finally told Joan that something in her personality prevented her from benefiting from therapy, she too was devastated.

There are a number of stories that describe how idealization contributes to the development of the person's own ideals. For example, in writing about his relationship to his older cousin, Richard stated that "I would try so hard to mimic him, trying to dress like him, act like him, trying to be him . . . he helped me grow in many different ways, he provided me with the brother that I never had, and that I was never able to find in my sister." And, of course, Jasmine described her father as her hero. Other women spoke of their fathers in similar terms; we have already seen how this was true for Ellen, who admired her father and felt supported by him.

Whereas the psychoanalytic theories emphasize that bonds with parents and others whom one idealizes help to shape the outcome of disillusionment, I found that the resolution of disillusionment typically involved support and help from people outside of these relationships. This was certainly also the case for the one person among Neubert's six interviewees who came to terms with midlife disillusionment and loss. The importance that others play in the healing process was evident as well in the study of self-forgiveness that my colleagues and I carried out.[44] Jasmine turned to a priest and retreat director (who reminded her of her father) and, through his intervention, she found a way to heal and to forgive her father. One might say that she brought her story of betrayal to the priest and, through this process, she was to some extent able to overcome the dichotomy (or "split," as Klein would say) between her father the hero and her father the man who abandoned her and her family. Benjamin also turned to a priest as he sought to come to terms with the end of his relationship with his girlfriend. Joan, who had been let down by her therapist, initially turned to a psychiatrist, as a kind of "higher power," to affirm her belief that she had been treated badly. Subsequently, she continued to work on coming to terms with this disillusionment through long-term psychotherapy. Margaret eventually went into therapy with her family and found that this both helped her to heal from the hurt of her stepfather's leaving and to develop a closer relationship with her brother. But the helpful other was not necessarily a professional. For example, Rosemary found support from her fellow campus ministry staff, and Ellen turned to her friends, who listened with care while avoiding the trap of siding with Ellen against her parents.

The differences in emphasis between my findings and those of psychoanalysis do not necessarily imply a contradiction. Thus, while Ellen was helped by her friends, it also seems reasonable to believe that the depth of her bond with her father was another aspect of her situation that allowed (or motivated) her to reach a resolution such that she could forgive her father and develop a new relationship with him. Psychoanalysis, traditionally, has had a deterministic bias—focusing in a one-sided way on how we are shaped by the past and, especially, by our relationships with parents or caretakers—and has given much less attention to peer and other nonfamily relationships. However, with the more contemporary thinkers in this tradition (such as Kohut), there is an increasing acknowledgment of the forward movement of human life, on how striving toward goals and a vision of one's own future gives direction to behavior and experience in the present. In Chapter 3, I discuss how, from a phenomenological point of view, the past, the present, and the future are interrelated. As T. S. Eliot wrote so eloquently:

> Time present and time past
> Are both perhaps present in time future,
> And time future contained in time past.[45]

The past, as lesson and memory, shapes the forward movement of our lives. Similarly, as our present attitudes and circumstances change, so do our recollections of the past and the lessons that we draw from them. With disillusionment there is a deep sense of loss and even of betrayal. Over time, insofar as we start to come to terms with what happened, we are more likely to view the person who lets us down as limited and fallible rather than as someone who is willfully destructive or unconcerned about us. This is similar to the movement that occurs in forgiveness, and we see it in several of the stories, most notably in Ellen's account of her changing relationship with her father, Margaret's story of how she is affected by the disappearance of her stepfather, and Joan's story of being let down by her psychotherapist.

In concluding this section, I want to address the creative response to disillusionment, a theme that encompasses possibilities for development of self structure (Kohut) or psychic structure (Jacobson), for a greater appreciation of life's precariousness and preciousness and a more realistic and reflective life philosophy (Janoff-Bullman, Klein, Holtz, Neubert, and Socarides), and for the creation of something new (Homans). We have seen how disillusionment throws people into a crisis and forces them to find their own way of making sense out of their lives and "to see things with their own eyes." As Carson Daniels emphasizes, those who achieve some degree of resolution become more conscious of how they need to be active in giving meaning to their own lives. No longer can one follow the path that an idealized other provides. Instead, one has to forge, painfully and actively, one's own direction. One young woman (in a story that I have not previously mentioned) describes this shift very powerfully:

> Looking back I can see how I became "disillusioned" by both my father and grandmother. Yet, growing older I have seen sides to both my father and grandmother that dismantle some of the painful beliefs I held about them during that trying time. I have learned rather to see them not through the eyes of that scared 12 year old but through the eyes of a woman who has learned from the past. The disillusionment I felt as a child has become "re-illusioned" as I begin to understand the complexity of the situation and the circumstances that both I and my father were under at the time. . . . In viewing my past in this way, and understanding my father at a personal level, I feel a greater empathy toward him than in the past.

Richard, who was betrayed by his the cousin when the latter took up with Richard's girlfriend, wrote in very similar terms of how he moved to a position of greater maturity: "Whenever I think back to those days and remember the grief that accompanied the disillusionment I felt towards my cousin, I have memories of sadness which one would think would be normal, but maybe more importantly I have memories of change that occurred, a change within myself, a change where I no longer had to view the world through another person's eyes." Both of these characterizations of change are very close to Jacobson's concept of development in psychic structure as including an increase in the person's capacity to manage his or her own emotional life, relate to others, take obstacles and opportunities into account, and respond to disappointment and conflicts.

Jacobson (and Freud) was alert to the possibility that the unrealistic belief in one's own specialness or the specialness of the person one admired could be transferred to one's ideals (or ego-ideal in psychoanalytic terms). In contrast, the disillusionment that is worked through instead can lead to a more sober and three-dimensional view of the world and others and to ideals that are more realistic. And, more generally, these statements also are reminiscent of the greater realism that Klein attributes to the depressive position as a way of patterning or organizing experience rather than as just a stage in childhood development. On the other hand, in several stories where there was much less of a sense of acceptance or resolution, there were hints of the problem of "transfer of specialness" about which Jacobson and Freud were concerned.

Self-examination is an important aspect of this process of maturation, and difficult questions about matters that were previously naively taken for granted are allowed to emerge. Holtz and Neubert would see this as a movement toward authenticity, and it necessarily involves a refashioning of one's assumptions as Janoff-Bullman suggests. Indeed, disillusionment and mourning can be occasion for creativity as Homans argued, but in this case not in the form of new theories or artwork but in the form of a new sense of oneself and the world and a different approach to relationships. This introspective turn is by no means a matter of turning away from others, especially not for those who do confront their own disappointment and ask themselves hard questions. One seeks out others for validation, consolation, and interpretation, but also with a keen awareness of the limits of what others can do to help. Briefly said, there is a sharper recognition of the separateness of self and other. Here again there is some similarity to the experience of "seeing the other" described in Chapter 1.

Conclusion

We have seen that in the center of disillusionment is the recognition that one has made the mistake, so called, of getting caught up in an illusion. How one then relates to and understands this mistake after the fact is of vital importance. In writing about working with clay pottery, Berenson provides us with a guiding image for coming to terms with the "failures" in our lives:

> In a very real sense, in this work, failure, if there is such a thing, can be viewed as a privilege. To make something successfully in the first place leaves you, in some cases, with little but a souvenir, whereas getting lost may afford you the opportunity to stop, examine and begin again and again with renewed insight and perhaps even personally developed solutions to move on with strength.[46]

This quote takes us back to Homan's emphasis on the importance of mourning and the possibility of a creative response to loss, as well as to Straus's suggestion that maturity requires a break with the past. To this many of us who have been disillusioned might well add, "I could not have done it alone."

But obviously there are many whose disillusionment is so great that they seem unable to come to terms with it, either on their own or with the help of others. It would be trite or unhelpful to suggest to those in such troubling circumstances that they regard this as an opportunity for learning or for growth, as if "reframing" the problem would readily give rise to a solution. What Socarides calls the "desire to remain disappointed," a self-protective stance of skepticism and cynicism that wards off the possibility of being disillusioned again by dismissing any sense of possibility for one's life, may well have become so deeply entrenched that the individual does not recognize it as a defensive posture. And even if there is such a recognition, it might make no practical difference insofar as the person has no lived sense of any viable alternative.

This is one reason why those who remain thoroughly disillusioned present such a challenge to their friends and family and to psychotherapists. As one relates to, or works with, a person who "remains disappointed," it might be helpful to keep in mind what Maurice Merleau-Ponty has said about a child's response to his or her life circumstances: "It is never simply the outside which molds him [sic]; it is he himself who takes a position in the face of external circumstances."[47] If this applies to the child, it certainly applies to the adult as well. To be mindful of how persons are more than victims of circumstances is to accord them respect in that one recognizes them as agents and to acknowledge that the solutions that they have crafted, however inadequate they seem to us, grew out of their attempt to repair their lives. And all of us are living with more or less inadequate solutions and awkwardly fashioned pottery. But there is a clear difference between personal solutions and artistic productions. Art objects, once made, endure on their own, while it takes energy for us to maintain our positions in the face of external circumstances, and so these positions are always apt to change, even if ever so subtly, as circumstances change and as we review our response to them. One of the things, ironically, that tends to strengthen a stance of disillusionment (and other similarly defensive stances) is direct external pressure, however well intentioned, to bring about change. Such pressure would include encouragement, in one form or another, to look on the "bright side" of life. One need not play into the hand of cynics either by going along with them or by minimizing the reality of disillusioning factors in the world. Instead, perhaps, one can gently (and indirectly) invite them to join in the common human struggle to find dignity in the face of disappointments and to develop ideals that are more sober and mature than the ones to which many of us initially gave our allegiance.

This discussion of the contributions of psychoanalysis to our understanding of idealization and disillusionment was meant to show the value of this tradition for helping us to look at a familiar landscape in a new way. For the most part, I believe, psychoanalysis does not contradict our everyday experience as much as it allows us the opportunity to think about it differently and to attend to subtleties and nuances that are readily or willfully ignored. The point is not that psychoanalysis is "right." Like phenomenology, it is rooted in experience—the psychoanalyst's experience of patients and his or her sense of their experience. Admittedly, this experience is reconstructed, sometimes quite elaborately, especially within the classical

Freudian schema. With the more recent theories, such as those of Kohut, the gap between observation and concepts is no longer so great. As I have indicated, some psychoanalysts use the term "experience near" to speak of ideas and concepts that have a more intimate connection with clinical observation.

In any case, we should not be afraid to see things with new eyes out of fear that theory will reduce or totalize our perception of ourselves, others, and the world around us. As long as we keep our eyes open, reality will disillusion us, again and again, of any conviction that any theory provides adequate images of the world of experience. There is nothing as practical as a good theory, but then we turn it into an ideology, just as we construct one-dimensional images of the people we know, mistaking them for the reality of who the other person is. The theologian Gregory Baum has warned us that "we can never be totally free from idolatry."[48] But perhaps, then, we can think of epiphanies and disillusionment as both, although in very rather different ways, freeing us (for the moment at least) from idolatry and ideology.

CHAPTER 3

Forgiving Another, Recovering One's Future

"Human forgiveness is not doing something but discovering something—that I am more like those who have hurt me than different from them."

John Patton[1]

"[Revenge] is dangerous, not because of what it does to your enemy, but because of what it does to you."

Laura Blumenfeld[2]

In the previous chapter, we saw how much our sense of self and our hope for the future are contingent on our relationships with those others who are an inspiration and a source of support in our lives. Whenever we are disillusioned by someone on whom we depend, our lives lose much of their meaning: we are no longer sure of who we are and the future becomes uncertain and unattractive. In the long run, we may come out of such crises with gains in the form of increased personal maturity and self-reliance. Yet, the more immediate consequences, such as grief, loss of confidence, and diminishment of the sense of one's value as a person, are deeply felt. The stories in Chapter 2 involved significant breaches of trust or failure to live up to what most of us would regard as reasonable expectations. Surely, it is not peculiar to think that therapists should act professionally and in their clients' best interest or that parents should care for their children.

Life is full of disappointments. Human beings are all too fallible, often failing even to live up to what most of us would regard as minimal expectations. In addition, as psychoanalysts have emphasized, our inclination to idealize others makes us vulnerable to being let down or induces us to engage in self-deception to avoid being disillusioned. It would be foolish to minimize any of these painful aspects of human relationships or to deny that for some the aftermath of disillusionment is a chronic attitude of bitterness and mistrust. On the other hand, we saw that in some of the stories betrayal and bitterness gave way to forgiveness and a new appreciation of one's value as a person. As part of such a resolution, the person who was disillusioned came to see the other person in much less personal terms and was able to appreciate what was valuable about the relationship prior to the disillusionment.

In this chapter, I take up the issue of forgiveness, briefly touched upon in Chapter 2, in greater detail. But forgiveness, as the quote from Patton suggests, also has connections back to Chapter 1 since at its core it also entails moving beyond a tightly held image of the other to the point of seeing him or her in a new and three-dimensional way.

Studying the Experience of Forgiveness: Beginnings

My own interest in this topic started about thirty years ago when I had an unexpected and unforgettable glimpse of what it means to forgive someone. It was a summer evening, and I was agonizing over a romantic relationship that had reached a painful conclusion, leaving me feeling devalued, hurt, and angry. To relieve my distress, I decided to go for a walk in my neighborhood. The sun had not yet set and so the colors of the trees and flowers were still vivid; in spite of being upset, I enjoyed their beauty as well as the peacefulness of the evening and the balmy air. I walked with no particular agenda in mind, and what happened next was completely unexpected. The following is part of what I wrote about this event several years after the fact:

> My anger and hurt vanished as I was thinking about Heather, but this time as another human being who was struggling, and who basically did not mean me any harm. It is not accurate, I am realizing, to suggest that I just thought that; it was more like an image that emerged for me, an image that was not as much seen as felt. I felt healed; blame and anger vanished, and there was a larger dimension of this whole experience that I can only describe in religious language: a sense of transcendence, of the future opening up, of a sense of presence, not of a personal being, but of connecting to something larger than myself and yet still having an experience of myself as me.[3]

Little did I know at that time that this incident would lead me to become involved in research on the psychology of forgiveness. This research has allowed me to see that my experience was similar to that of many others and to situate what happened within a larger context of understanding.

A few years later I was fortunate enough to participate in a National Endowment for the Humanities Summer Seminar with Professor Robert

N. Wilson of the University of North Carolina at Chapel Hill. The seminar, which provided a wonderful venue for discussion with colleagues, had, as its focus, sociology, psychology, and literature. The readings included several of Eugene O'Neill's plays. I decided to base my presentation for the seminar on O'Neill's understanding of forgiveness as presented in his last two plays, *Long Day's Journey into Night* and *A Moon for the Misbegotten*. These plays address the difficulty of moving toward forgiveness within families and are closely based on the painful and conflicted history of the O'Neill family.

In 1977, the year of the seminar, there was very little literature in psychology on forgiveness, leading me to conclude that "psychology has treated forgiveness with benign neglect."[4] The reasons for such neglect of what is obviously a profoundly important topic lie in the nature of the discipline of psychology itself. Historically, psychology sought to free itself from its connection to religion and philosophy and to establish itself as a scientific discipline. Textbooks in the field proclaim that psychology began as an autonomous discipline in 1879, the year that the physiologist and philosopher Wilhelm Wundt opened a *laboratory* in Leipzig, Germany, for the study of psychological phenomena. As Amedeo Giorgi has pointed out in his critique of mainstream psychology, this discipline has based its claim to being a science on the fact that it relies on the experimental method, just as do the natural sciences, especially chemistry and physics.[5] Consequently, psychology has tended to ignore phenomena that could not easily be studied using standard research methods.

It is far from obvious how one could adequately study forgiveness using the experimental method. Unlike aggression or persuasion, it is not a phenomenon that can readily be "produced" in the psychological laboratory. In contrast, an experimenter can quite easily evoke anger or aggression in experimental subjects by subjecting them to conditions of frustration or deprivation. Moreover, psychological theories are hard pressed to account for forgiveness. As the political philosopher Hannah Arendt has pointed out, forgiveness overcomes the tyranny of the past by providing us with a new beginning, by opening up the future.[6] Although my experience of forgiveness was not the end of the hurt and conflict, it was the beginning of a process that moved me toward such freedom. Social scientists, including psychologists, tend to think in terms of determinism, the assumption that our actions in the present are the result of what happened to us in the past. The way in which forgiveness overcomes the effect of the past by changing its meaning for the present (and future) is at odds with this assumption. As a result, psychology has, until recently, been content to let theology and religion take care of the topic of forgiveness.

It is just in the last decade or so that there has been a steady increase in the number of psychological studies on forgiveness and, particularly, on how one can help people who have been injured to forgive. In some ways this change is encouraging. At last psychologists are studying this phenomenon that is so important in a world so fraught with strife and injury. Paradoxically, though, in these studies, little attention is given to people's actual experiencing of forgiveness. Or, to put it differently, psychologists are more interested in their own *theories of,* and *procedures for,* bringing about forgiveness than in what forgiveness involves in everyday life. The result is that one now finds in the psychological literature numerous assertions about the nature of forgiveness, a number of which are unsupported, untrue, or misleading. This is a truly peculiar state of affairs, which I will discuss later in this chapter. But first I want to look closely at what descriptions of the experience of moving toward forgiveness can tell us about the process itself.

In late 1984 my colleague Jan Rowe and I began a series of studies on forgiveness, first addressing forgiving another and then self-forgiveness.[7] These studies were carried out by small collaborative groups, which included four to six graduate students as researchers. We did in-depth interviews to get detailed first-hand accounts of the participants' struggles with injury and their journey toward forgiveness. We also wrote out accounts of our own experience with forgiveness and analyzed these descriptions along with the stories of the people we interviewed. The basic questions that we asked of our interviewees (and of ourselves) were: "Can you tell us about the time during an important relationship when something happened such that forgiving the other became an issue for you?" and "Can you describe a time in your life when self-forgiveness became an issue?" Doing this research gave us critical insights into the nature of forgiveness, helping us to see what makes it so difficult and what makes it possible. Much of the material in this chapter is based on these studies.

In addition to participating in this series of studies, I have been teaching a seminar on the psychology of forgiveness at Seattle University for over a decade. Each year I have asked students to write descriptions in response to the same two questions that we developed for our research studies. These descriptions and the discussions with the students in these seminars have also deepened my understanding of this topic. However, I want to emphasize that forgiveness is an elusive and intricate phenomenon; there is always more to it than I or any other researcher can say about it. This is one reason why I am including several stories about forgiveness in this chapter and want to look closely at them. The other reason is that traditional psychological studies have neglected first-hand accounts. These stories that follow are distinctive and yet they have much in common. All of them point to the mystery that is inherent in human life, to the depth of persons, and to the way in which our existence unfolds in unexpected ways.

Learning from Stories of Forgiveness

I start with a story that, because of its clarity and vividness, is especially effective in exemplifying the key features in the forgiveness process. In this case, the person who is forgiven is a stranger. Subsequently, I will discuss another situation where the forgiveness process involves two people who are intimately involved, and then I will refer to a third story where the person who is forgiven has been dead for a number of years.

Valerie Fortney, who is a professional writer, published her story in *Chatelaine,* a Canadian magazine, in 1997.[8] At the beginning of the story she is twenty-nine. One September evening, just before going to bed, Fortney watches a news report about an accident caused by a drunk driver. Half an hour after she has fallen asleep, a phone call awakens her. It is her sister who calls to tell her that their other sister, Shelley, has been killed in that very accident, and that Shelley's boyfriend has been severely injured. This terrible loss brings to her mind the snowmobile accident sixteen years earlier in which their brother was killed. The family is told by the police that the driver of the car that went through a red light and hit the car carrying Shelley and her boyfriend was drunk. He fled the scene, but returned later with his wife and gave himself up.

A week after the funeral, she sees the man who killed her sister when he appears in court and pleads not guilty. He is a very ordinary-looking man in his early thirties, who stands with his hands clasped and looks at the floor. Two weeks later, Fortney, her sister, and her mother are at a shopping mall, and she sees him again, this time with his wife and son, apparently enjoying themselves. She goes up to him and tells him who they are. He looks stunned and quickly disappears during the moment when Fortney looks back at her family.

Fortney finds out about the man's background and the drinking party he attended that preceded the accident. She becomes obsessed with him and has constant thoughts of taking revenge by running him down with her car or shooting him. At the same time, she hates herself for having such vengeful thoughts. Her anger and preoccupation take over her life and she becomes increasingly reclusive, drinks and eats excessively to deal with her pain, and gets excruciating headaches. She is caught up in a self-destructive pattern that continues for a number of months. Eventually, there are signs of a shift in Fortney life. Around Christmas, she drives by the man's home, thinking of going in to hurt him, but ends up just sitting in her car and weeping.

Six months after seeing the man at the shopping center, Fortney runs into him during a lunch break at the preliminary court inquiry. Having in her mind the excuses his lawyer gave, such as that her sister's boyfriend drove recklessly, she walks up to him with a sense of strength, and says, "What gets me the most is that you don't even have the basic human decency to say that you are sorry." Much to her surprise, the man folds over as if someone has hit him in the stomach, and blurts out, "But you don't know how sorry I am."[9] He then breaks into heart-wrenching sobs. She is shocked to realize that she feels an urge to comfort him, that she feels compassion for him. Instead, she talks about her sister's life and about the death of her brother.

This is a critical turning point for Fortney. As she puts it, "The discovery that this man whom I'd demonized was a feeling fallible human being gave me the release I'd been looking for." She starts to feel sadness, for herself and the two families involved in this tragedy. Her obsessions gradually go away as does her anger at the world, and she begins to go on with the living of her own life. Now, when she thinks of her sister, she can celebrate her life and who she was rather than being focused on her death. When the driver is sentenced to thirty months in jail, after changing his plea to guilty, she does not have a sense of victory but a deep sense that everyone has lost with this tragedy.

The issue of forgiveness arises in a wide variety of relationships and situations. But on the basis of our research, we have concluded that there are some fundamental dimensions to the experience, no matter what the context.[10] First, when people speak of forgiveness they begin with a detailed description of a profound injury (or injuries). The injury is such that the world as once known is dramatically disrupted—turned upside down in some sense: there is a wrenching and tearing of the very fabric of one's existence. The future as imagined or taken for granted is no longer possible, and one's assumptions about the other(s) are called into question. One's assumptions about oneself are also called into question or undermined and bring such thoughts as "Why me? What have I done?" One is uncomfortable in one's own skin, ill at ease with oneself. The world, literally the very street one lives on, is experienced as somehow alien, untrustworthy, perhaps foreboding. In Fortney case, the death of her sister completely changes her life. It is as if she no longer has a future; all of her living revolves around the fundamental loss and injustice that she has experienced.

Emotionally, one's response involves intense rage, sadness, and profound distress. One's energies are focused on the injury and on the person who has committed the wrong. For Fortney, as for many others, rage and thought of revenge predominate at first, and grief only gradually becomes present. When one finds oneself momentarily distracted from the hurt, even a small reminder triggers the distress—waking up in the morning (if one has slept) and a recollection puts one back in the midst of one's anger/sadness/anxiety/confusion. This preoccupation with the hurt, or "licking one's wound," has two dimensions: (1) the immediate obvious experience of the injury and (2) a deeper and larger meaning for one's life. The first level of injury is the centre of one's attention at the time; the latter may become clear later. At the first level, Fortney is devastated by the death of a member of her family, as one would expect. The second level is harder to articulate since she does not tell us directly what this loss means to her. We might surmise that the death of her sister tells her that the world is an unjust place, that there is no safety for her or her family (she has already lost her brother in an accident), and that there is no way that she can protect herself or those she cares about.

Another example may help to further clarify the distinction between these two levels of injury. Let us suppose that a man is told by the woman to whom he is engaged that she has decided not to marry him because she loves someone else. This announcement is likely to be experienced as being very hurtful. It means that his relationship with this woman is over and that they will not get married. He may also feel betrayed. At a deeper and more metaphorical level, he is likely to wonder whether there is something wrong with him and whether he will ever find anyone who will love him. That is, what he is told by his fiancée takes on implications that extend beyond his relationship to her. What she says to him shatters his image of who he is and what the world is like.

One experiences one's self as "injured," "wronged," "diminished," and "vulnerable." Attempts to cope with the powerlessness of the situation may involve fantasies of, and desire for, revenge, and fantasies and acts of self-destruction, for example, excessive drinking and withdrawal (as was the case for Fortney) and, in some cases, thoughts of suicide. Yet, however alone one may feel in one's pain, the role of others is important. Having people who are supportive, who allow one to experience whatever is happening without judgment, and who, by their very presence, affirm the seriousness of what has occurred can be very useful as one tries to make sense of the situation. Fortney says little about her contact with friends except to comment that she became reclusive and that the whole episode put a strain on her relationships. Nonetheless, she continued to have contact with her family, who shared in her loss.

At some point, realizing the harm one is inflicting on oneself, one may want things to be different. Fortney wrote of hating the person she was becoming. Yet a way through to some other place is not clear. One may even attempt to make this happen, telling oneself, "I'll just put this behind me," "I need to get on with my life." But these attempts are short-lived at best. One may feel relatively free of the hurt one day, only to find oneself enraged all over again the next day. However, these periodic breaks from the pain are significant, in that, while one may yet feel stuck, they still offer a glimpse of a life beyond hurt and rage.

Especially when the injury is in the context of an ongoing relationship, one may begin to reflect on one's own part in the injury that led to the wounding. More generally, one may wonder about the other and who this person is beyond merely being a perpetrator, trying to imagine the situation from his or her point of view. When Fortney drove by the man's house at Christmas, there was a sense that she wanted to know more of who he was. And although she went with thoughts of hurting him, she ended up sitting in front of his house, weeping. However gradual or dramatic such a shift in perspective is, it involves a loosening of the rigidity of the grip of one's distress.

Even though one wants to move forward, there is also a resistance to the movement—a resistance to letting go, as if something precious would be given up. In some ways the resistance to forgiveness seems to involve a fear of what letting go would entail: the deep sadness and grief about what could have been had the injury not occurred, full acknowledgement that things will never be the same, relinquishing of the hurt and rage that in some way have come to be part of one's identity, and the facing of an unknown future. So there is a battle within oneself—between a desire to move on and a fear of what that would mean. This struggle seems to be a necessary part of the movement toward forgiveness, even if it is a difficult process with which to be contending.

When forgiveness happens, it typically comes as a surprise. Just as Fortney described how her perception of the man who killed her sister had changed quite unexpectedly, forgiveness is experienced as something that has happened rather than something one has caused to happen, and the experience is profound and transformative. As defining as the injury and its impact were, so is the experience of forgiving. People report feeling lighter, fuller, clearer; whereas, before, the future was foreboding, it is now full of possibilities. There is a sense of freedom to move into the future no longer burdened or limited by the hurt. While the injury is not forgotten, it no longer holds one so tenaciously. People describe feeling reconnected or connected in a new way to others. There is a fresh awareness of the humanness of all persons—including the wrongdoer—and it is this awareness of the profound humanity of oneself and others that brings the deeper meaning of forgiveness into focus. Lastly, because forgiveness is not experienced as an act of will, it has a transcendent dimension. This is reflected in the words people use to describe it, such as "an act of grace," "the opening of a door," or "a gift"—something more than the person's doing, with unimaginable benefits. Fortney is quite surprised to find herself celebrating her sister's life and what her sister meant to her, and being able to go on living her own life.

The second story is about the relationship between a man and a woman who are intimately involved. I had two lengthy interviews with "Dave" as part of our study of forgiveness.[11] His story brings into clear focus critical aspects of forgiveness while also showing that it is an extended and multifaceted process. Dave spoke of "small forgivenesses leading up to a final forgiveness." In what follows, I have tried to do justice to what Dave told me, even while summarizing his story.

A college student in his late twenties, Dave had been living with "Sarah" for about four years. About a year and a half before the interview, Sarah developed a friendship with "Gardner," a colleague at work. Initially, Dave did not object to this relationship because he assumed that Sarah placed the same priority on having a relationship based on the ideal of commitment as did he. Also, since he was very involved with his studies and his part-time work, he was glad that Sarah had a companion during those times he was busy. When Dave discovered that Sarah and Gardner's relationship had become a romantic and sexual one, he asked Sarah to decide between him and Gardner. When she seemed unable to decide, he became all the more insistent. The ambiguity of the situation became intolerable for him. For Dave the hardest part of this situation was not Sarah's sexual infidelity but that he was living with someone who said she loved him while also apparently loving someone else. In this situation, Dave felt jealousy, resentment, and hostility. After almost a year of tensions and momentary resolutions that never lasted—Sarah said she would stop her involvement with Gardner, but did not, and said she would move out on her own, but stayed—Dave took the initiative to call Gardner and request that the three of them meet to discuss this situation. Gardner reluctantly agreed to the meeting.

While speaking to him on the phone, Dave realized that Gardner was anxious and uncertain, and this realization brought about a shift in Dave's attitude. Rather than continuing to think of Gardner as an enemy and as "morally inferior," Dave understood that the two of them were in a similar situation. He said, "It was like my heart went out to him, I felt very compassionate, and that was when I think . . . that was kind of when I forgave him." In the following excerpt from the transcript, Dave described the further developments that took place when the three of them met face-to-face a few days later:

> The thing that really crystallized the experience for me, and my ability to forgive, was that I sort of met him face to face as a human being who

wasn't a threat or someone stealing something from me, but I realized that he was someone with the same needs and that . . . we were all in this conflict together . . . afterwards, ever since then, it was like I was able to completely, without trying, to let go of it . . . but the important thing was that I was able to forgive myself for demanding certain things, certain a priori values out of the relationship, such as total commitment, things like that. I was able to let go of Sarah's indecisiveness, and I was also able to let go of the idea of Gardner as some kind of an interloper or enemy, and so the forgiveness for me wasn't just a granting pardon to these two people but it was that the whole thing just sort of slipped away, it didn't have the same emotional charge, and it was sort of a peacefulness that's been there, and an acceptance of the situation which no longer seems difficult or threatening.[12]

Subsequent to his meeting, Dave felt much more comfortable with Sarah's relationship with Gardner as he no longer looked to her to fulfill his own needs. About a month later Sarah moved out to live on her own. Dave and Sarah continued to see each other, and their relationship improved.

There was another incident that enabled them to reach a deeper level of forgiveness. About a month after Sarah had moved out, Dave had stayed overnight with another woman to whom he was attracted, and Sarah found out about it the next day. She was terribly hurt and angry and asked Dave how he could do this to her. His response was that this is what she had been doing to him for a year and a half. During this exchange, they became increasingly emotional and upset with each other. After some time, they reached the point where they were able to share their deep grief at what had happened in their relationship and to comfort each other. Eventually, they came to what Dave described as "the final resolution" of the hurt they had caused each other. In speaking of the change that all of this brought about in terms of his relationship with Sarah, Dave said, "I really see her as another person instead of somehow an appendage of mine, so when we are together, when we make love, I feel like here's this other person I am coming together with; this is a privilege and a beautiful thing."

Here, once more, we see how forgiveness is a movement of compassion. Throughout the two interviews, Dave described how he came to recognize in a most immediate way how the people whom he came to forgive—Sarah and Gardner—were people like him, contending with the same issues and difficulties as he was. In regard to Gardner, he commented, "Before he had been like a bad feeling, someone ethically inferior, a pawn of Sarah, and now my heart went out to him."[13] The other whom one forgives, as Patton writes, is someone like oneself. Compassion acquires significance or becomes real, Kierkegaard suggests, only when the compassionate person identifies himself with the sufferer, when that person's situation is directly related to one's own.[14] And Milo Milburn, in his phenomenological study of forgiving another, concludes that empathic identification with the person who injured oneself is an essential aspect of forgiveness. By this he means "the finding of the Other in oneself and of the self in the Other, of a common, wounded, fallible yet valued humanity."[15]

In elaborating upon the phrase "my heart went out to the other," Dave spoke of his whole being focusing upon the other and of deeply appreciating Gardner as a separate individual. So compassion involves a paradoxical movement of letting go of one's preconception of the other, connecting with the other as similar to oneself and yet being aware of the other's separateness. This awareness of separateness is exemplified by Dave's relationship with Sarah after this resolution. He stated that he no longer looked at Sarah as the person whose responsibility it was to help him with the struggles in his own life.

Compassion toward the other is a core dimension in the experience of forgiveness, but it need not happen in a face-to-face situation; it can also happen in imagination, as my description showed, and, in the case of one woman I interviewed, in a dream. This woman dreamed that she embraced her son as he was dying. Her son had committed suicide two years previously, leaving her completely bereft. She had never had a chance to say goodbye to him. When she woke up from the dream, she was crying, and she found that the world was restored to her in the sense that she could now again enjoy life, even as she continued to live with sadness.[16]

Forgiving another (and oneself) involves taking responsibility for one's own contribution to the painful or problematic situation. For example, Dave explained how he had become aware of having shifted the burden of responsibility for making a decision to Sarah, having neglected her through his involvement with work and school and having simply assumed that his affirmation of the principle of commitment was the same as an active ongoing commitment. What taking responsibility actually means varies in critical ways depending on the circumstances of the injury. Obviously, Fortney was not in any way responsible for the death of her sister, but she had to recognize and take ownership of how her reaction to this tragedy was destroying her own life.

Without some awareness of one's own fallibility, it would be impossible to connect compassionately with the other as someone like oneself. However, taking responsibility for oneself should not be understood in a moralistic sense, with a connotation of self-blame or accusation. It is best described as an owning up to, or embracing all of, who one is and has been, one's life, and one's actions. The steps leading up to this shift, may, nonetheless, be extremely difficult and fraught with pain as one agonizes over one's own actions and limitations.

Letting go is clearly part of forgiving as the conventional wisdom suggests. But what is it that we let go of? Dave frequently spoke of letting go of expectations—another word for demands. More specifically for him, forgiveness meant "becoming attuned to yourself and your experience, and letting go of your expectations, for one thing . . . and of the desire to make the other person yours or mine." In regard to Sarah, he said, "I trusted her to be what she was instead of trusting her to be what I wanted her to be."[17] So, letting go is not just a negative movement. Rather, it involves reconnecting with life at a deeper level, while, at the same time, being freed from being caught up in, and feeling responsible for, others. In a very profound sense, letting go is an appropriation of freedom and responsibility, of one's own finitude, and of hope.[18]

In hope, Gabriel Marcel reminds us, one no longer identifies one's own well-being or fulfillment with a particular fate.[19] Herein, in part, is the critical thread in the movement from despair to freedom. As one becomes open to the future as possibility (the essence of hope), time is no longer a prison from which one cannot escape unless a particular event occurs. This opening to the future is evident in Dave's case. He spoke specifically of having gained a greater awareness of his own autonomy and needs and of realizing that it was his responsibility to take care of these needs rather than expecting someone else to do so. It was no longer a matter of his fate depending upon Sarah's decision. And Fortney was no longer so exclusively preoccupied with the death of her sister. Instead, she was able to grieve, to remember her sister with fondness, and to continue on with her own life.

What allows us to let go? First, we need to keep in mind that our anger and indignation at others is not in response to what they do per se, but to the effect that their actions have on us, our position in life, and our relationship to self and others. Their effect on us depends, as we have seen, in part on our own agendas and expectations. Taking responsibility for our own contribution to the situation is part of the forgiveness process, and this is part of what Dave did. He also lessened his insistence that the future yield to his demands. During the meeting he even suggested to Gardner that perhaps he was the better partner for Sarah. Gradually, then, he found himself less caught up in how particular people responded to him; his position in life was not so directly a function of what Sarah decided or did not decide.

Included in this process of letting go is the acceptance of one's own finitude and limitations. Perhaps one of the things that made it easier for Dave to move toward this kind of acceptance was that he had been confronted precisely with this issue five years before this incident. At that time, he had struggled to accept his dependence on others and his own mortality during a bout with a potentially fatal illness. In dealing with this illness, Dave had to come to terms with the difference between the life that he wanted and the life he had, between the person he thought he was and the person that he actually was. He could not take it for granted that he would have a long life or even that he could manage the details of his everyday existence without help.

Forgiveness entails a letting be, a letting go, and therefore it is not to be construed as an act of the will any more than the creation of selfhood is. In fact, Dave spoke of the belief of being in control through willpower as an utter sham. Every one of the people we interviewed for our research projects indicated that they *found* that they had forgiven someone, or had forgiven themselves, and this was also, as we have seen, the case for Fortney. Consequently, we characterized forgiveness as a "gift," a word that was used by several of the people we interviewed.

Of course, this is not to suggest that forgiveness happens against one's will. A gift has to be received, and letting go is a willing act, although not a willful one. Leslie Farber, a psychiatrist influenced by Martin Buber, has perceptively distinguished between willing and willfulness, or what he refers to as the first and second realms of will, in his book *The Ways of the Will*.[20] This distinction is very important for an understanding of human experience, but it is rarely touched upon or acknowledged in psychology.

In discussing the first realm of will, Farber is referring to an underlying, even unconscious, willingness to move in a particular direction. Thus, "will itself is not a matter of experience, though its presence may be retrospectively inferred after this realm has given way."[21] That is, after the fact, can we look back and see that we were open to or heading in some direction, without having had a specific agenda or goal in mind? It was not as if we deliberately or consciously made a decision to do something in particular or that we were intently pursuing a specific outcome. In both of the stories given earlier, as well as in my description, we can see that there was a willingness or receptiveness of which the person was not directly aware. Fortney goes to the house of the man who had killed her sister, originally with the thought of hurting him. Instead, she ends up sitting in front of his house, crying. As she grieves, there is a lessening of rage and hatred. That there is movement, even though not yet clearly discernible to her, is also evident in her response to this same man when he starts sobbing after she accuses him of being indifferent to causing the death of her sister. As he cries, she is surprised by her inclination to reach out to him. It is easy to imagine someone else, still consumed by hatred and anger, who would not be moved by such a display of regret and guilt. In other words, the man's expression of regret does not by itself account for her response; she had already moved to a position of being more open to him than previously.

Similarly, we can think of Dave's urge to call up Gardner and suggest that the three of them get together to discuss the awkward and painful situation that they were in. Dave did not call up Gardner with any conscious thought of forgiving him. But just as Fortney was profoundly touched when the man who had killed her sister started sobbing, so too was Dave moved by what he perceived as Gardner's vulnerability. And, likewise, the experience that I had of forgiving Heather was a completely surprising but nonetheless welcome "gift." Along the same lines, as we saw in Chapter 1, allowing oneself to be surprised by the other involves a degree of openness and responsiveness.

Think of your own experience of being with someone who was open or willing. More specifically, bring to mind a time when a friend listened carefully to what you wanted to tell him, and you felt like you had his undivided attention. Then, imagine what would have happened if you told this person that you appreciated his openness to you and what you said to him (perhaps you actually said something like this). It is not likely that he would have said, "Yes, I know that I am open." More often, he would have responded, "Really?" as if he had not thought of himself in this way. In fact, he was not thinking of himself at all—he was listening to you. Similarly, think of a time when you were really open or receptive to a situation or a person. Were you consciously aware, at the time, that you were so receptive, or did you realize it only after the fact? In contrast, recall a time when you were determined to be open-minded. Did your determination bring about genuine openness, or did you realize, again after the fact, that telling yourself that you were receptive, or that you should be open, did not make you so?

The second realm of will, the practical will, is a matter of a state of mind that we experience directly. Farber is referring to a deliberate or utilitarian kind of striving, an explicit setting of goals, an implementation of activities, which are within our conscious control. We speak of someone as being determined or intent, and, if this determination becomes overextended, as being willful. Within the realm of the practical will, our attention is quite narrowly focused on a limited aspect of a situation that is directly related to our agenda. Thus, in a conversation, we may seem to be conversing quite casually, but in reality there is a specific goal that we are pursuing. I might want to know if the other person likes me, or if I can sell something to him or her. The examples Farber uses are especially helpful in distinguishing between the two realms: "I can will knowledge but not wisdom; going to bed but not sleeping; eating but not hunger; meekness but not humility; scrupulosity but not virtue."[22]

This leads to the question: "Can you will forgiveness?" The answer provided by the stories given in this chapter and by the findings of our research is, "No, you cannot." Indeed, the determination to will forgiveness creates problems, including self-recrimination, insofar as one recognizes the failure to do so, and confusion and self-deception, insofar as one thinks one has forgiven but in reality has not. Patton, drawing upon his extensive experience as a pastoral counselor, has discussed the false sense of power that comes from the belief that one can grant forgiveness. That is, if I convince myself that it is entirely within my power to forgive the person who has wronged me, then I am like a king who has the power to pardon his subjects if he is so inclined. In that case, one can set conditions for granting forgiveness to the other person. I might insist, for instance, that the individual express regret or ask for forgiveness. Ironically, if the person does express regret, I might then realize that I am not able to forgive.

Patton further suggests that if we have been injured, we might take refuge not just in this illusion of power but also in a sense of righteousness. I have been wronged; the other person is the wrongdoer. I am in the right; the other person is in the wrong. These beliefs, Patton argues, protect me from the sense of vulnerability, impotence, and shame that come with being injured. In the short run such responses make sense, and the sense of righteousness may have some justification. However, in the long run, these ways of protecting oneself may bring about additional problems insofar as one becomes mired in them and they become part of one's identity. Thus, they become obstacles to forgiveness and to living a life that is not defined primarily by one's injury.

This discussion may create the impression that I believe one's conscious intentions are not relevant, and that choice has no place in forgiveness. This is not what I am suggesting. However, as the Canadian philosopher H. J. N. Horsbrugh has argued, "The decision to forgive is normally only the beginning of a process of forgiveness that may take a considerable time to complete."[23] Yet even this is an overstatement of the role of decision if one thinks of it as a primarily conscious act. Did Fortney and Dave decide to forgive? Not in the ordinary sense in which we use this word. But, obviously, they gradually reached the point where they were *willing* to forgive. This position is also confirmed by Milburn's study of forgiveness. He found that for the people he interviewed, the movement in the direction of forgiveness preceded any explicit decision that they wanted to forgive.[24]

Before I return to the topic of contemporary psychology's treatment of forgiveness, I want to refer briefly to another remarkable story about forgiveness. John Douglas Marshall's story *Reconciliation Road: A Family Odyssey of War and Honor* is distinctive in two ways.[25] First, it provides a detailed account of the road to forgiveness and is told by the person who experienced this journey. It is important to keep in mind that people's actual struggles around an issue of forgiveness are far more complex, convoluted, and difficult than any summary, such as that I have provided, can adequately convey. Second, this is the story of a man who forgives someone who has been dead for a number of years. This account provides comfort for those who assume that forgiveness is only possible if one has direct contact with the person by whom he or she was wronged. The students in my course on forgiveness read this book and discuss it with the author at the end of the quarter. Many of them have found the reading and the discussion beneficial.

My intention here is not to summarize the story but to briefly discuss some of its key features. I would encourage readers to read the full story themselves.

John Marshall's grandfather, S. L. A. Marshall, was a Brigadier general in the U.S. Army and a famous war historian. John was very close to his grandfather as he grew up, and had great admiration for his work as a writer. Their relationship changed drastically and painfully when John became a conscientious objector during the time of the Vietnam War. John sent a detailed and eloquent letter to his grandfather, explaining his reasons for taking this step, even though he was in the Reserve Officer's Training Corps. He was apprehensive about how his grandfather would react; the letter he received in response was more devastating than anything he had expected. Essentially, his grandfather called him a coward and "excommunicated" him from the family. John was stunned and hurt. He never saw his grandfather again and did not go to his funeral.

Twelve years after his grandfather's death, when John Marshall was a reporter for the *Seattle Post-Intelligencer,* a controversy erupted about whether his grandfather had been a reliable historian or, as some claimed, had made up his facts along the way. After some deliberation, John decided, with the support of his wife, to set out on a journey (September to December 1989) to check out this controversy as objectively as he could. His road trip took him to the University of Texas at El Paso, where the Marshall archives were held, and to dozens of other places, including Washington, D.C., as he met with and interviewed people who had known his grandfather or who knew of his work. These people included Mike Wallace of *Sixty Minutes* and General William C. Westmoreland. On this journey, he spent time with his father and his siblings and visited his mother's and his grandfather's grave. All of this changed John Marshall, bringing him closer to several members of his family and to a new understanding of his grandfather's vulnerability, foibles, and strengths. Unexpectedly, his journey brought about a measure of healing for himself. At the end of the trip, John Marshall wrote the following about his relationship with his grandfather:

> We were so different. Vietnam brought that out. What I have found on the road has only reinforced that. . . . Still, this journey, born amid my lingering anger and ambivalence about my grandfather, has brought me to the point where I can honestly say I forgive him for what he did to me. I still deplore it, but I forgive him now. Finally.[26]

John Marshall regretted that forgiveness did not come about while his grandfather was still alive, and he mourned for the years of pain and estrangement. Nonetheless, something fundamental had changed. He was now able to speak freely to his son about his grandfather and to think about him without being so troubled by the letter that had been so hurtful.

It is often said that we cannot change the past, and in one sense this is true. But in another it is not. John Marshall did not set out with any conscious intention to forgive his grandfather or to change the past. However, by the end of his trip, both had happened. He had come to the point that he took the letter from his grandfather much less personally, understanding how much it was an expression of his grandfather's fallibility and touchiness. The letter continued to be a fact of history, but its significance had changed. It is the significance of the past that shapes our present and our anticipation of our future, not the "facts in themselves" (as if such facts exist). Maurice Merleau-Ponty, the French philosopher, has written eloquently about the relationship between the past, the present, and the future: "Each present reasserts the presence of the whole past which it supplants, and anticipates that of all that is to come, and by definition the present is not shut up within itself, but transcends itself towards a future and a past."[27] In other words, what happens in the present affects our relationship with the past and with the future.

The other stories of forgiveness in this chapter also show how the meaning of events can change. I came to understand Heather's behavior as reflecting self-protectiveness rather than meanness; Fortney came to view the death of her sister as a tragedy for everyone involved rather than an attack on her and her family; and Dave came to appreciate that he was not just an innocent bystander, victimized by Sarah's indecisiveness and infidelity. In each of these situations, the past became less defining, the present more open, and the future more promising.

Contemporary Psychological Approaches to Forgiveness

We have seen how much stories of forgiveness can tell us about this phenomenon. Phenomenological psychology, to put it very simply, takes descriptions and reflects on them with the aim of arriving at an understanding of what is essential or basic to these descriptions. The key question is, "What is true across descriptions beyond what is distinctive about each one?" In Chapter 5, I will say more about the underlying assumptions of phenomenological psychology as well as the steps that are involved in its practice. For the moment, I will look at how mainstream psychology has approached the topic of forgiveness.

Psychologists started to pay attention to forgiveness in the early 1990s, and there is now a substantial body of literature on the topic. I do not intend to do a systematic survey of this literature. Rather, I am going to use a few selected examples to show how psychology characteristically deals with forgiveness and to discuss value and limitations of this literature.

Robert Enright, in the Department of Educational Psychology at the University of Wisconsin at Madison, has probably done more than anyone to make forgiveness a respectable topic of study within mainstream psychology. Starting in 1985, Enright and a group of graduate students and faculty began a seminar on forgiveness. They were concerned with determining what forgiveness is and is not and with devising a model that describes how a person moves toward forgiving another.[28]

To arrive at an understanding of the nature of forgiveness, Enright and his colleagues studied religious and philosophical texts, as well as contemporary writings on forgiveness in psychology, psychiatry, and religion. They were also strongly influenced by the English philosopher Joanna North. In a 1987 article North asserted that "forgiveness is a matter of a *willed* [italics in original] change of heart, the successful result of an active endeavor to replace bad thoughts with good, bitterness and anger with compassion and affection."[29] The Enright group's definition—which changes a bit depending upon at which one of the publications one looks at—is explicitly based on North's ideas. The following is a recent version:

> People, upon rationally determining that they have been unfairly treated, forgive when they willfully abandon resentment and related responses (to which they have a right), and endeavor to respond to the wrongdoer based on the

moral principle of beneficence, which may include compassion, uncondi-
tional worth, generosity, and moral love (to which the wrongdoer, by nature
of the hurtful act or acts, has no right).[30]

The stories I have presented, as well as the overall findings that came out
of our research at Seattle University, also confirm that to forgive is to move
from resentment to compassion. But the language in this definition, especially the emphasis on rationality and will, is puzzling. Those who forgive
tell us that they *have* forgiven, and this discovery (as Patton calls it) is not
necessarily, or even typically, preceded by a conscious decision to do so.
This, of course, does not mean that they forgave in spite of themselves, but
it does suggest that forgiving is more a matter of being willing than of being
willful. Even in those cases where the people decide that they would like to
forgive, the decision does not by itself bring about forgiveness, anymore
than a decision to quit smoking results in abstinence. How can one account
for this gap between Enright and his colleagues' definition and what people
report? The answer, quite simply, is that the definition is arrived at theoretically rather than empirically (empirical means derived from experience or
based on the facts). North states that her orientation is based on the philosophy of Immanuel Kant (1724–1804) and others who belong to the
"rationalist school." This school believes in the central role of rational
thought in human life, and Kant in particular believed in the capacity of
will not just to control actions but also one's emotional responses.[31] Few
contemporary observers of human behavior (be they historians, journalists,
or psychologists) would find this view compelling.

This definition also implies that the person who forgives stands above,
in a moral sense, the person who is the wrongdoer and bestows something
upon him or her. Although there are certainly situations in which one
person is clearly wronged by another (as was the case for Fortney), many
situations of injury are more ambiguous. More fundamentally, the compassionate stance that is part of the process of forgiveness implicitly involves
an embracing of the other as a fellow human being. To their credit, Enright
and his associates recognize that while (in their view) forgiveness is something one chooses, it also results in a discovery, insofar as "to forgive means
to begin seeing the other in a new way, as a member of the human community rather than as evil incarnate."[32]

In my view, the members of the Enright group (following North's lead)
are more successful in their endeavor to distinguish forgiveness from other
phenomena with which the concept is sometimes confused than they are
in their attempt to define it. Forgiveness, they argue, is different from reconciliation in that one may or may not renew the relationship with the

person one forgives. Most notably, a woman who has been abused may eventually forgive the man who abused her, but by no means does it follow that she will decide to live with him again. Nor is forgiveness the same as pardoning. Fortney forgave the man who killed her sister, but he still went to jail. The judge had no reason to pardon him; after all, he was guilty as he himself acknowledged.

Enright and his colleagues have developed a model of forgiveness that identifies the phases involved. They list four phases (uncovering, decision, work, and deepening), each of which is divided into units. For example, the uncovering phase includes "units" such as confronting anger and admitting to shame, the work phase includes "reframing" who the wrongdoer is (that is, taking a different perspective on this person) and giving a moral gift to the offender, and the decision phase includes willingness to consider forgiveness as an option and commitment to forgiving the offender. The idea of the model, at least in theory, is that the phases and units are sequential. If you have not confronted your anger, for instance, then you cannot move on to let go of that anger. At the same time, Enright and his associates recognize that the forgiveness process is not so linear, and that any model is necessarily a simplification. There are aspects of their model that, in my view, are problematic, and these problems are similar to the ones in their definition. However, setting aside whatever limitations the model may have, it has evidently provided the basis for several systematic studies of interventions to help people move toward forgiveness.

Perhaps the most noteworthy among these studies was the one carried out by Suzanne Freedman while she was one of Enright's graduate students. It led to a 1996 publication with Enright as the second author. The study's goal was to demonstrate that incest survivors could be helped to heal through forgiving their abusers.[33] Given how severe a violation incest is, the study's goal was certainly ambitious. The study was structured, as is typical in mainstream psychology, as an experiment where one group of people were given the "forgiveness treatment" and the other group served as the control group. Freedman recruited twelve incest survivors using ads and public notices (these mentioned healing but not forgiveness). Half of the volunteers were at random assigned to get the treatment first; when this group was done, the control group became the second experimental group. As a result, everyone had the opportunity to benefit from the treatment. The people in the experimental group were given a manual that described the phases in the Enright process model, and Freedman then met individually with the survivors once a week for an average of fourteen months. She worked with them individually, making it possible for each subject to move through the units (e.g., admission of shame about the incest) at her own pace.

Change in the participants was measured using a variety of psychological scales that assessed self-esteem, depression, anxiety, hope, and forgiveness. These scales were given before the treatment, a second time when the subject had reached the point of saying that they had forgiven the abuser, and then a year later. Overall, there was a significant improvement in these women's psychological well-being as they moved through the study. They became less depressed and anxious, and more hopeful and forgiving (as measured by a scale and according to their own reports).

In their discussion of the results, Freedman and Enright address the question of whether the improvements are due to the skill of the therapist or the power of the specific intervention. They take the very defensible position that in practice one cannot separate these two factors. It is unlikely, they argue, that someone who did not believe in the value of this approach to forgiveness or did not have adequate therapeutic training could be very effective with it. But Freedman and Enright also point out that the results gained were quite remarkable compared with the generally limited effectiveness of most forms of psychotherapy with incest survivors.

Rarely do articles in mainstream journals provide qualitative or descriptive data in addition to the quantitative data. Freedman and Enright are an exception in that they include an appendix, entitled "Case Study Illustrating Forgiveness Process," in their publication. Although the case study is written from the researchers' perspective, it does give us some indication of one person's process of forgiveness. "Nicole" was fifty-one years old and had been sexually abused by her father when she was a child. In addition to this trauma, she had also had cancer. Her life, by anyone's standards, had been extremely traumatic. She had been in psychotherapy for eight years. One of the interesting points in the story is how she came to finally forgive her father. Nicole had gradually gained a better understanding of her father by learning a lot about his background and by remembering some of his good points. At a point when she had not yet forgiven him, although she had come to a better understanding of the source of his disturbance, she sent him a card and a gift on his birthday. In turn, she received two letters from her father, which helped her to reach the point of forgiving him. These letters showed her the more positive and vulnerable side of her father. This story shows how multilayered the forgiveness process is and how it involves not just one's own actions and decisions, but also one's relationships and interactions with others.

Although the number of participants (or "people") was small, this study (along with others) provides evidence that it is possible to help people move toward forgiveness, even in the case of those who have been severely wronged. However, the value of the Enright process model in providing the basis for this kind of intervention does not by itself confirm that its steps and underlying assumptions are "correct." As I have already pointed out, some aspects of Freedman and Enright's theoretically based model are contradicted by the findings of studies that are based on descriptive data. In addition, as we know from other areas in psychology and psychiatry, a variety of interventions may be similarly effective even though they are based on very different understandings of the problem being addressed. The treatment of depression is one of the clearest examples of this paradox.

The biological approach to depression assumes that it is caused, at least in part, by the decreased availability of the hormones serotonin or norepinephrine in the brain. Medication in the form of antidepressants has been shown to be effective, at least in a majority of cases, in reducing depression. There is evidence, although it is not clear-cut, that the medication increases the production of serotonin and norepinephrine. Cognitive models of depression argue that negative thought patterns (e.g., focusing on mistakes and discounting successes) and irrational assumptions ("everyone should like me") predispose a person to depression. Cognitive therapy, as developed, for example, by Aaron Beck, starts by gently guiding patients to recognize, in very specific terms, how their continually bleak perception of the world and themselves stems from their own negative thinking patterns. The second step is to guide the patients toward a more realistic form of interpretation and thinking. Again, there is strong evidence that cognitive treatment is effective even with serious depression, even though it is not clear that negative thoughts are a primary cause of depression. Finally, interpersonal therapy, which is indebted to interpersonal-oriented psychoanalytic theory (such as that of Harry Stack Sullivan), assumes that depressed people have unproductive ways of relating to other people, tending toward either avoidance or "clinging," and have difficulty processing grief. The therapy focuses on analyzing the patients' current relationships and helping them to interact more effectively. It too has a good track record. So we can see that all three forms of treatment have been shown to have considerable value even though they are based on largely contradictory views of depression.[34]

Overall, psychologists discuss forgiveness in rather reductive terms. By "reductive," I mean that this process, which is subtle and profound, is frequently described in ways that are simplistic and one dimensional. For example, Michael McCullough and Everett L. Worthington, two psychologists who have written a great deal about forgiveness, call it a "religious behavior," hail it as a "promising therapeutic tool," "and assume that it is something that a person who is injured "grants" to the individual who injured him or her.[35] Enright and Richard Fitzgibbon's approach is more nuanced in that they affirm that forgiveness involves transformation and that it is more than a therapeutic procedure. Yet they also describe it as an action and encourage clinicians to "consider the use of forgiveness" as if it were a medication or technique.[36]

We are faced, again, with the limitations of psychology insofar as it restricts itself to following methods developed within the natural sciences. At the same time, the truth is that few psychologists have any training in doing qualitative research. Furthermore, the prevailing ethos within their discipline inclines them to be skeptical about the value and the validity of descriptive accounts. First-hand accounts are typically dismissed as "merely anecdotal," which is another way of saying they are not trustworthy. The psychiatrist Elio Frattaroli has pointedly taken to task this distrust of human experience endemic to psychiatry and psychology. "Why should we need a laboratory experiment to convince ourselves of something we can know directly from our own inner experience? The idea that we do —that no knowledge can be considered scientific, valid, or true unless it has been proved in a controlled laboratory experiment—is a modern aberration, a radical misconception about scientific knowledge that may serve ideological prejudices (and managed care-business practices) but clearly does not serve the pursuit of truth."[37]

There is one last issue that I will touch upon here and discuss further in the next chapter. When psychologists refer to forgiveness as a technique or to evidence that certain procedures are effective, they demonstrate how much reliance on, and faith in, techniques as a vehicle for progress and for reaching preconceived goals has become a feature not just of Western society but of the disciplines of psychology and psychiatry. What is meant by technique, in this context? William Barrett, in *The Illusion of Technique*, has defined a technique as a standard method that can be taught. Further, he specifies that it can be taught because its steps can be precisely specified.[38] The steps could be the ones the psychologist uses to help a person move toward forgiveness or the ones that the subjects or clients are taught so that they can attain control over their emotions and behavior and help themselves toward the same goal.

The assumption is made, at least by some in the psychological community, that techniques can be applied to human affairs in a way that is very similar to how they are employed in relationship to the world of objects and of nature. A further assumption is that techniques provide primary solutions to human problems; that is, they allow us to escape issues such as ambiguity, choice, and interpretation. These assumptions add up to a rationalistic and reductionistic view of human nature.

Why does this tendency toward reductionism matter so much, and what are the consequences of dismissing first-hand account? In other words, what is at stake here? I turn to one last story to answer these questions and to bring this chapter to a conclusion.

Empathy for One's Enemy

In 1985 Brian Keenan moved from Dublin to Lebanon to teach literature at the University of Beirut. This was a dangerous time for foreigners to be living in Beirut. A number of Americans and Europeans had been kidnapped by various terrorist groups that operated with relative impunity during this time of political violence and social instability. The kidnappers hoped that by holding hostages they would have leverage to influence Western governments' policies toward Lebanon and other countries in the Middle East or to bring about the release of their imprisoned compatriots. Although he was from Northern Ireland, hardly a major player in the region, Keenan was kidnapped not long after arriving in Beirut. He was held hostage for four-and-a-half years in conditions that were physically and psychologically abusive. His book, *An Evil Cradling,* is an eloquent account of those years of suffering and of his own resourcefulness in dealing with his imprisonment.[39]

Keenan had the company of a fellow hostage, John McCarthy, with whom he became friends. They were lorded over by unpredictable and often brutal guards and were kept in a small room, with limited access to toilet facilities, with chains on their ankles and feet, and with little news of the outside world. At times they were threatened with death and on a number of occasions they were beaten.

One of the guards was especially brutal. Said was a small man, who ranted and raved, and who went to great lengths to humiliate any of the hostages with whom he came into contact. The hostages knew that he had lost his wife in a car bomb explosion, but the viciousness of his conduct pushed aside any sympathy that they might otherwise have felt toward him. Indeed, McCarthy and Keenan dealt with their rage toward him by discussing their fantasies of overpowering and castrating him.

A sheet was hung across the room in which the hostages were kept, dividing the space into two. McCarthy and Keenan were on one side, and the guards were on the other side where the window was. The guards occupied their time by talking, watching television, as well as listening to tapes of a holy man who was chanting and reciting parts of the Koran. Said would listen to these tapes and start chanting along, eventually working himself into a delirious state. The hostages would be driven to distraction by this noise and what Keenan described as "grief-stricken hysteria."

One day, when Said was there alone with Keenan and McCarthy, he again worked himself into a state of agitation while listening to the radio. While his fellow hostage was dozing, McCarthy found himself listening to Said, as the latter was moving around restlessly. But then his restlessness gave way to a deep sobbing, unlike anything Keenan had previously heard from him. Something remarkable then happened as this hostage took in what he heard of his guard's suffering:

> I felt, as I never had before, great pity for this man and I felt if I could I would reach out and touch him. I knew instinctively some of the pain and loss and longing that he suddenly felt himself overwhelmed by.
>
> The weeping continued, Said became fleshy and human for me. Here was a man truly stressed. His tears now wrenched a great well-spring of compassion from me. I wanted to nurse and to console him. I felt no anger and that defensive laughter which had before cocooned me was no longer in me.[40]

This momentary experience is remarkable. It says something about the human capacity for compassion and for transcendence, of seeing another and relating to another, if only for a brief period, in a new way. Perhaps it is a glimpse of forgiveness, but whatever one might want to call this transformation of Keenan's view of Said, it points to the way in which human beings are fundamentally connected through their vulnerability and their grief. I think it is no exaggeration to say that this kind of experience highlights the limitations of psychological theory and reminds us of how human experience and behavior are rooted in relationships and in history, how it is always in context.

To some extent, we can relate what happens here back to psychological theories about the nature of forgiveness. The concept of "reframing" has some relevance to Keenan's description. It is one of the critical units in the Enright formulation of the phases of forgiveness, and is basically a process whereby the inured person deliberately attempts to see the wrongdoer within the context of the person's life in order to get a more complete picture of the wrongdoer as a whole person and of his or her actions.[41] But this concept and others like it are not sufficient. The very experiences to which these concepts refer ought to be examined on their own terms because the concepts are too narrowly focused on the individual and because they reduce the depth of life to something that can be fully explained and managed. However, life is not something that happens in the psyche or mind of the person but in his or her concrete and ever-changing relationships with others and the world. In the conclusion to our study of forgiving another, we wrote, "The experience of forgiving the person who has injured oneself is a complex multidimensional process that moves from a tearing of one's lived world through feelings of hurt, anger, revenge, confusion to an opening up to a larger experience of oneself and the world."[42] The disappointing aspect of psychology's increased attention to this topic is that the discipline has not allowed itself to be opened up to a broader understanding of human behavior in the process of studying forgiveness, a topic that is so richly revealing of the depth of human existence. As I have tried to make clear, part of this resistance is rooted in psychology's almost exclusive reliance on the experimental method as the royal road to understanding. This is all the more reason to remember how vital it is to learn from the accounts of those who have struggled with injury and forgiveness. For, despite the depth that they reveal, these stories have, at best, only a marginal place in psychological research. And yet, I would claim, there is no higher authority on the topic of forgiveness.

CHAPTER 4

Experiencing the Humanity of the Disturbed Person

I become through my relation to the *Thou;* as I become *I,* I say *Thou.* All real living is meeting.

Martin Buber[1]

We, as a society, are *estranged* from the "mad" in our midst. We fear them and their illness.

Robert Whitaker[2]

Before I started writing this chapter, one of my colleagues asked me why I was including a discussion on the mentally ill in a book on interpersonal relations. It was a good question, and one for which I did not immediately have an answer. This is not, after all, a book on abnormal psychology. Yet I intuitively felt that this inclusion not only made sense but that it was critical. Accordingly, I begin this chapter by explaining why understanding people with psychiatric disturbances, in addition to being an important topic in its own right, has significant implications for ordinary human relations.

There are three basic points that I want to develop in what follows. First, coming to understand someone who is mentally ill happens in fundamentally the same way as with anyone else who initially puzzles or confounds us. In either case, one arrives at an awareness of the point of view of the other person, seeing something of how he or she experiences a given situation. This is, in a nutshell, the core of empathy. By considering what makes empathy possible in this specialized context, that is, in relation to the mentally ill, we can gain a fresh perspective on the process of empathy as it occurs more generally. In this way, the current chapter connects back to Chapter 1.

Second, our appreciation of our own humanity is deepened insofar as we find a way to recognize the humanity of someone else who initially appears to be very different from us, whether on the basis of psychiatric disturbance, race, illness (e.g., AIDS), religious affiliation, nationality, or age. David Kahn, a professor of nursing who studied the lives of residents in a nursing home, writes poignantly of how his relationship to these elderly people changed during the time he carried out his research:

> This experience [of fieldwork] has immersed me in ways and understandings of life that were foreign to me. In retrospect, it is painfully obvious that I was guilty . . . of beginning this work with a view of the very old as objects somehow apart from my own understanding of human subjectivity, despite my clinical experience with them (or maybe because of my clinical experience). This has changed over the past months as I listened to them talk about their lives. I have given little to them other than some company and an opportunity to reminisce and pass time. They have given me far more. Through their words, I have glimpsed the essence of my own old age as I carry it now in what and who I care about in this world, and what I will some day lose.[3]

There is certainly a bias in our society toward the mentally ill, just as there is toward the elderly. We avoid the mentally ill because they reflect back to us possibilities that we fear in ourselves. This is not to imply that we are, without consciously knowing it, hovering on the edge of madness! Rather, I think it is fair to say that few of us are either familiar or comfortable with many of our own supposedly irrational and uncontrollable inclinations, thoughts, and emotions. As a result we shy away from situations that bring any of these aspects of ourselves to mind.

Finally, theories of psychiatry and abnormal psychology influence our view of human nature, and not just of "abnormality." We are fascinated by "abnormality" because it is a scary and intriguing arena, and, simultaneously, we seek explanations that might clear up the sense of mystery surrounding this issue. Currently, most such theories are reductionistic at the same time that they speak with the authority of science. In addition—and this is often overlooked—they seek to explain human behavior in general. Taking these explanations to heart leads us to think about human nature, and thus about ourselves, in ways that do not do justice to the complexity and richness of our own humanity. Thus, we should acknowledge that these theories are influential, often of therapeutic value, and yet also one-dimensional. For example, some address just the biological or the cognitive aspect of the person.

I want to begin with what may seem to be an absurd assertion: the mentally ill are disappearing. They are disappearing because of the degree to which reductionism characterizes contemporary psychiatric as well as popular thinking. Increasingly, we do not as much see as *see through* the mentally ill. This is true even while they are far from disappearing from public view. If anything, the contrary is the case, as many mentally ill people live on the streets and as the media give significant attention to mental health issues. In what follows, I first discuss how the mentally ill are increasingly present among us. Then I briefly explain my assertion that they are, nonetheless, also disappearing. This is followed by a discussion of how one can understand the mentally ill from a psychological perspective. Sadly enough, such a perspective, which looks at the mentally ill in more personal terms, is currently being overshadowed by biomedical interpretations. Overall, my aim is to outline a positive and humanistic vision of psychiatric disturbance. With this vision as background, I critique what I regard as depersonalizing trends in psychiatry and psychology.

The Increased Presence of the Mentally Ill

During the last forty years the number of mentally ill people living outside hospitals has risen dramatically. Between the 1950s and the 1990s the number of mentally ill people in hospitals in the United States decreased by four hundred thousand.[4] During the 1960s and 1970s a larger number of people with psychiatric diagnoses were discharged from hospitals, sometimes after careful preparation, but more frequently with little planning and with minimal provisions for follow-up and support in the community. Sometimes this shift is described as a program of "deinstitutionalization." In contrast, Gerald Grob, who has specialized in the study of changes in the treatment of the mentally ill, argues that this massive discharge of patients was not the result of a program but the consequence of a series of largely unrelated factors.[5] Regardless of how this pattern came about, it involved monumental change.

Before the 1950s, those who were admitted to state psychiatric hospitals were often kept there for a long time, and many were never discharged. Patients were typically not sent home until they showed significant improvement. Since there was little or no effective treatment available in these institutions, it is not surprising that patients seldom met this criterion. A few private hospitals provided intensive treatment. But because these hospitals were costly and had only a few beds, they were outside the reach of the vast majority of people needing treatment.

Several factors contributed to the shift from hospitalization to deinstitutionalization. A number of sociological studies have shown that the longer people stayed in hospitals, the more they became institutionalized. That is, they adjusted to living as patients. Institutional routines do little to prepare people for living successfully on the outside. As a patient, one does not take the bus, drive a car, cook a meal, go to a grocery store, apply for a job, plan a trip, look after children or aging parents, buy a home, or rent an apartment. One lives according to someone else's dictates (however well intentioned), one is identified as disabled, and one's basic needs are automatically provided for. Altogether, it became apparent that hospitals did not prepare patients for returning to the community.

Of course, these studies by themselves did not bring about reform. However, their publication coincided with a number of powerful changes in society and psychiatry. There was an increased optimism during this period that people with mental illness could learn to cope with living in the community. The advent of major tranquilizers (or neuroleptics) in the 1950s made it possible to reduce some of the most evident and severe symptoms of mental illness, such as hallucinations and delusions. By the 1960s, the civil rights movement, with its emphasis on liberty and opportunity for all human beings, was an important social force. This was also a time when humanistic psychology was quite influential. This perspective emphasized the value and dignity of each person and opposed involuntary hospitalization, diagnosis, and any form of coercive treatment. Keeping people shut up in mental hospitals did not fit well with the values espoused by either the civil rights movement or humanistic psychology.

Economic considerations, as one might expect, also played a major role. It is very expensive to keep people in hospitals, and if hospitalization characteristically does more harm than good, then the money is wasted. Moreover, by 1972, the Social Security Act was modified so that those who were mentally ill became eligible for support, and so, at least in theory, they would have the means to live in the community.[6] Tragically, little of the money saved by discharging patients was put into community mental health programs that many believed might allow these same patients to live reasonable lives outside of hospitals. Essential services such as psychiatric assessment, medication, psychotherapy, training, and rehabilitation were in short supply. The Community Mental Health Act of 1963, passed during the Kennedy administration, did provide some federal funding for community mental health, but without strong state support this was not enough. As one would expect, funding for mental health fluctuated according to who was in the White House. The Carter administration was supportive of mental health programs, while the Reagan administration sought to cut funding for such programs. In any case, it is all too apparent that mental health services have been, and continue to be, a low priority in most states. It is no wonder that at least a third of the homeless are people with mental illness. More recently, after the terrorist attacks of September 11, 2001, the funding for mental health was decreased yet further as most state governments were faced with severe budgetary shortfalls. Tragically, the trend continues. In Oregon, for example, a significant number of mentally ill people living in the community lost funding for psychiatric medication. Since the cost of this medication may run as high as $500–600 per month, their situation became untenable. Very few of these people can pay for medication on their own, and it is entirely predictable that without the medication to which they have become accustomed, many will become disturbed and require hospitalization.[7]

But it would be a mistake to think that most of those people in the United States with a psychiatric diagnosis have been hospitalized. According to the surgeon general's 1999 report on mental health, 20 percent of all Americans have at least one mental illness in any given year.[8] This means that an even greater percentage have at least one mental illness over a lifetime. Some of these psychiatric disturbances may be quite "mild," in the sense that they do not prevent the person from working and living a relatively normal life. People with acrophobia (intense fear of heights), for example, are not directly troubled by their problem except in situations involving heights (flying, taking an elevator), and there are many people diagnosed with depression who take medication and are thereby able to modulate their mood swings. However, there are certainly also those whose disturbances, even if they do not result in hospitalization, prevent them from working or developing the kinds of ordinary relationships that most of us take for granted. Mental illness (a very imprecise term at best) includes a diversity of disorders and can be applied to many of us, at least at certain points in our lives.

It is generally agreed that more people today are diagnosed with mental illness—or would receive such a diagnosis if they were evaluated by a mental health professional—than was the case thirty years ago. There are several explanations for this increase. Certain disturbances are more prevalent than in the past. Clinical depression, as the most dramatic example, may be about ten times more common in the United States today as it was around 1900.[9] On the other hand, schizophrenia, another major disorder, is roughly at the same level as it was a hundred years ago. No one is able to explain why depression has increased so much while the incidence of schizophrenia has remained constant, although it is likely that multiple factors—social, psychological, and biomedical—are involved.

The second explanation focuses on the criteria used in diagnosing mental illness. Not only have the criteria changed over time but new diagnoses have been added. The diagnostic manual published in 1968 by the American Psychiatric Association (APA), the *Diagnostic and Statistical Manual of*

Mental Disorders (DSM-II), had less than one hundred disorders. The subsequent versions, published in 1980, 1987, and 1994, list over two hundred.

Is the increase in the number of diagnostic categories a problem? One could argue that disturbances that were previously unrecognized and thus untreated have finally been given the attention they deserve, and that treatments can now be developed. One might also wonder, though, if too many human problems are being turned into medical issues, thus pathologizing more and more people. For example, a recent addition to the list of psychiatric diagnoses is "caffeine intoxication."

Additionally, some evidence suggests that life has become more stressful and chaotic for a number of people. I have spoken to a number of clinicians who have worked in mental health centers and similar agencies for a long time. They report that their clients have become increasingly disturbed over time as the hardships that they deal with on a daily basis have intensified: extreme deprivation and poverty, parents who are on drugs, violence in the neighborhood and in the family, lack of community support, and lack of opportunities for adequate education or reasonable employment.

The Vanishing of the Mentally Ill

It seems then that the mentally ill are with us more than ever, and that, in a manner of speaking, the mentally ill are us! Of course, as I have already mentioned, Sigmund Freud long ago proclaimed that we are all more or less neurotic, but he did not have the authority of the medical establishment to back up his assertion. If, however, psychiatric problems are so widespread, then why am I suggesting that the mentally ill are disappearing? My claim is that they are disappearing in the sense that there is decreasing emphasis on understanding the mentally ill *as persons*. Admittedly, there is nothing new about the mentally ill being overlooked or, in the not-so-distant past, literally being hidden away in asylums or hospitals. There is a vast body of literature on the brutal treatment of the mentally ill throughout history and on prejudice and discrimination toward them in the current age. The surgeon general's recent report on mental health argues that stigma toward the mentally ill has increased in the last forty years largely because of the erroneous association of psychiatric problems with violence.[10]

Ironically, the current lack of emphasis on understanding individuals with psychiatric diagnoses as persons is partly a function of recent innovations in the study, treatment, and diagnosis of mental illness. These developments include advances in neuroscience, such as increased sophistication in the methods available for the study of brain functioning (e.g., computerized tomography and magnetic resonance imaging), clearer diagnostic criteria, and the promotion of standardized methods of treatment through the use of manuals for therapists. It is as if those with psychiatric problems have been moved outside of the realm of the interpersonal as their problems and lives are being examined less and less from a psychological perspective as medically based and impersonal approaches take center stage. The problem is not that there have been advances in biomedical approaches. Rather, the problem is that little attention has been given to the limitations and one-sidedness of these perspectives. These limitations include the fact that they deal with the organism rather than with the person. In addition, because there have not been commensurate advances in humanistic approaches to persons with psychiatric disturbances, the biomedical advances are looked upon uncritically.

This chapter, then, takes us back to the Introduction and the theme of the disappearance and appearance of persons. As we recall, in G. K. Chesterton's story, Professor Openshaw failed to notice that Mr. Berridge was a fellow human being, not just a clerk who took orders from him. Not until he was shocked into awareness did the professor recognize his clerk as someone who had a mind and a life of his own. But what does it mean to understand people with disturbances as fellow human beings? In ordinary discourse, we describe these people as "insane," or "out of their mind," with the implication that their behavior, unlike like our own, makes no sense, that it is irrational. We assume that their actions are caused by mental illness and that, therefore, there is no need for further explanation. Thus, essentially, we are denying that those who are disturbed have a story to tell to which we can relate. As we will see, this position is not just one held by lay people with little background in psychology. It is also held by a number of people in mental health. In order to explain why I think this stance is mistaken, and to show how insight into the life of someone who is mentally ill can come about, I will tell some stories of my own.

Understanding Persons who are Disturbed

My first exposure to the field of mental health left a deep impression on me. After my sophomore year at Glendon College of York University in Toronto, Canada, two fellow students and I volunteered to spend a summer living in the maximum security section of the Penetanguishene Psychiatric Hospital, north of Toronto. The Oak Ridge section of the hospital was designed to house the "criminally insane." These patients had a diagnosis of serious mental illness (such as schizophrenia, "psychopathy," or profound depression) and had been judged by the court to be "innocent by reason of insanity." This means that although many of them had committed a serious crime, there was strong evidence that they were so seriously disturbed at the time that they could not tell right from wrong or, did not, in any meaningful sense, understand what they were doing. Dr. Elliott Barker, the psychiatrist who was in charge of the Oak Ridge section, had been working for about a year on developing an intensive treatment program in one of the wards. The patients on "G" ward were volunteers from the other seven wards who had agreed to participate in a therapeutic community program.

This approach to treatment, based on a model developed in England by Dr. Maxwell Jones, emphasized active participation by patients in their own treatment and in the treatment of fellow patients. Buber's notion of the healing power of dialogue was also an important influence on the program. An early paper on the rationale behind the program was entitled "Buber behind Bars."[11] The program had a strong social and psychological orientation and was based on the belief that patients could make headway in dealing with their psychiatric disturbance by gradually taking on positions of responsibility and by being supported as well as confronted by peers and staff. The patients had agreed to the three of us becoming temporary residents on their ward, but they probably had as much trepidation about our stay as we did. They were anxious as to how they would be evaluated by three "sane" people. In turn, we were mindful of the fact that about a third of the patients had committed homicide, worried about whether they would accept us, and wondered how well we would cope with living in a prison hospital. Our reasons for taking part in this experiment were varied and complex. I was a history and political science major and was considering changing my major to psychology and therefore wanted to learn more about mental health and psychotherapy. Indeed, this seemed like a great opportunity to gain first-hand knowledge about a world that was largely unknown to me. No doubt I was also struggling with my own demons, including questions about my own sanity. At that time, I was both moody and introverted.

Dr. Barker believed that it would be beneficial for the patients, many of whom had been hospitalized for years, to have an opportunity to interact with ordinary people. The goal of the program was to enable at least some of the patients to improve sufficiently so that they could be sent back into the community. Since no one had yet been released, Dr. Barker and the other staff also wanted to see what it would be like for someone (in this case the three of us) to make the transition from the highly intensive hospital program to the outside world.

All three of us were profoundly affected by our stay, even though it only lasted two months. The emphasis on expression of thoughts and feelings within the program and the constant interaction with our fellow residents as well as staff resulted in our becoming more emotionally expressive, often stating rather directly exactly what we were feeling. It took some time for us to readjust to living outside of the hospital. Everyday social conversation seemed superficial, and we were initially overwhelmed by the number of decisions required of anyone living "on the outside." I remember panicking the first time I went to a restaurant and was confronted with a menu with a seemingly endless array of items from which to choose.

Taking note of our posthospital turmoil, the staff concluded that there needed to be a structured transitional program for anyone who was going to be discharged. But the main point I want to make in telling this story is that we developed relationships with the patients that were relatively uncluttered by psychological theory since none of us had much background in this discipline. Although some of the patients stayed aloof, others were eager to talk to us. We certainly came to see them as fellow human beings, however disturbed they might have been and however appalling the actions that had brought them to this hospital were. As we lived with these men twenty-four hours a day, we caught glimpses of their concerns, fears, and hopes, their yearnings to be loved and to live a meaningful life. This was truly an unforgettable summer, one that changed my life and my vocational aspirations. By the time I returned to school in the fall, I had decided to become a psychotherapist and accordingly changed my major to psychology and sociology.

The following summer I returned to Penetanguishene, this time as an assistant to Dr. Barker. In that position, I read the psychological evaluations of a number of men on the unit. And although I did not doubt that these reports had value, I was struck by how abstract they were—how little connection there seemed to be between my knowledge of the same patients as human beings and how they were portrayed in these evaluations. I was left wondering how such a gap could be overcome.

The beginning of an answer emerged before I started graduate school in psychology. I spent half a year as a child-care worker at a residential treatment center for disturbed children and adolescents in a small town just outside of Windsor, Ontario. The treatment center, with residents ranging in age from six to eighteen, was located in a large house.

One of the oldest and most disturbed residents was a seventeen year old whom I will call John. John had been transferred to this treatment center from a psychiatric hospital where he had been given a diagnosis of paranoid schizophrenia. Schizophrenia, a serious mental illness, is generally regarded as being unresponsive to psychological treatment. John was fortunate to have been transferred to us. We did not assume that he was a hopeless case. Whatever the limitations of our program, he was assured of personal attention and plenty of opportunities for interaction with the staff and fellow residents.

There was no doubt that John was troubled. Much of the time he seemed intensely uncomfortable in his own skin and he had less skill in interacting socially than did his fellow residents. He often muttered to himself and kept talking about how his life would end in 1972 (this was in 1967). When he became agitated, John showed signs of what is technically called "thought disorder," one of the classic symptoms of schizophrenia. Thought disorder involves using words in unconventional ways or making up one's own words, moving from one idea to another without establishing a connection, and speaking in such a way that the casual observer concludes that the words are just thrown together (hence the term "word salad" is used to describe some forms of thought disorder). As an example, in response to a question about the meaning of the proverb "Rolling stones gather no moss," a patient might say, "I know of no moss and I do not want to stone any rolls, so let me be free to gather leaves." The listener, of course, is left completely baffled as to what this means. Thus, thought disorder is one of the psychiatric symptoms that gives credence to the belief that the mentally ill are beyond understanding. When John became agitated, he went on and on in this manner, visibly angry about something we did not understand and unable to explain what was going on for him. Sometimes, when he was upset, he would retreat to his room and play his favorite record over and over again. Not surprisingly, the other residents tended either to provoke him (this was easy to do) or to avoid him.

One day John came with me when I ran a number of errands with the agency station wagon. We spent several hours together, driving from one location to another, dropping off laundry, picking up groceries, and so on. John was relatively relaxed and spoke quite personally about his life and the deep disappointments he had experienced. For example, it became clear to me from our conversation that being liked by the other residents was very important to him and that he was deeply distressed by his lack of progress in this regard. I saw something of how painful his life was.

We returned in the late afternoon and started to carry laundry and groceries from the car into the house. John entered the house just ahead of me. He was barely inside the door before Frank, one of the other adolescent boys, demanded that John go into the kitchen to get him a Coke. John did so immediately, with an eagerness that reminded me of how important approval was to him. Unfortunately, John made a tragic mistake! When he returned from the kitchen, he handed a Pepsi to Frank. Frank became very angry and swore at him. The change in John was immediate and dramatic. He became agitated, started to pace back and forth and yell incoherently about how his life would end life in 1972 and the other topics that he rambled on about whenever he was upset. But this time I heard all of this in a new way. Having just listened to him for several hours, I now understood most of what he was saying, notwithstanding the presence of "thought disorder." I understood how terribly hurt John was, how the hurt was more than he could tolerate, and how his "delusion" that he would die in 1972 made his life more tolerable. That is, I grasped that in the midst of a life that was deeply distressing, the belief that his life would be short provided some solace. And I saw again what I had seen with a young patient when I was at the Penetanguishene Hospital: behind the apparent craziness was a tremendous vulnerability.

Some years later, I had a similar experience while working in a state psychiatric hospital in Pennsylvania.[12] Dorothy was a stocky, vigorous woman in her early forties who was hospitalized with a diagnosis of bipolar disorder. She had both manic episodes where she appeared to be completely out of control, sometimes throwing furniture across the day room on her unit, and periods of depression when she felt hopeless and showed little initiative. Those who worked with her had been unable to determine what brought about her changes in mood. Yet one thing was clear: her relationship with her husband was vital to her. There was little else to give her life meaning. The couple did not have any children, and Dorothy did not work outside of home. Once we developed a treatment plan with her and her husband where the reward for her improvement was that she would be able to go home for visits on weekends, with discharge from the hospital as the long-term goal, Dorothy became quite cooperative. On the surface, things were going well. However, our concern was that while Dorothy was eager to go home, her husband was less enthusiastic about this goal.

One day I was at the front office when the mail arrived. There was a letter for Dorothy and I decided to bring it to her since I was going up to her floor. When I found her, she was calm and coherent as she had been all week. She noted that the letter was from her husband and took it to her room to read; I went to the nurses' station across the hall to write in patients' charts.

A few moments later Dorothy came charging out of her room and threw the letter on the floor. Her face was flushed, and she ran down the hall clearly agitated, singing and rambling incoherently. Soon all the staff on the unit knew that Dorothy had become manic again. Moreover, they assumed that the disturbance had come out of the blue. However, when I read the letter that she had left behind, it became obvious to me that this was not true. On the surface, the letter did not seem to warrant such a reaction. Her husband simply referred to some of the difficulties that she would have in living outside of the hospital. Yet, having met this man on several occasions, I had realized that it was necessary to read between the lines to understand what he was really saying. The conclusion that I drew from his letter, and the one that I assumed Dorothy had also come to, was that he really did not want her to come home. But coming home was the one goal she had been working toward, and the one possibility that would give her life meaning. Without it, her life had no purpose, no direction.

The staff who wrongly assumed that the change in Dorothy was unrelated to her circumstances had also made two other unwarranted assumptions. The first was that a diagnosis explains behavior: Why did Dorothy become manic? Because she is bipolar and that is what patients with that diagnosis do. Their second assumption was that because they could not see any reason for her reaction, there was no reason. On the other hand, I understood her behavior differently because I, both literally and figuratively, was given the opportunity to look over her shoulder by reading her letter and recognizing something of its implications for her life. Being a witness to these circumstances and knowing something of the context of her life, I knew that her manic episode was by no means out of the blue.

Understanding people's behavior requires us to grasp the connection between what they do and the situation in which they find themselves. In social psychology (in contrast to much of psychiatry), the central tenet is that people's immediate circumstances shape their behavior. As James Waller has written, "Mainstream social psychology has long believed that what really matters is not who you are, but where you are."[13] In his work he has tried to account for the fact that ordinary people have committed horrendous acts of evil by relating their behavior to a variety of situational factors.

This emphasis on a situational analysis within social psychology is valuable, but it is not sufficient because the other half of the equation is that where you are depends, in some way, on who you are. For example, when it rains, just about all of us open our umbrellas, suggesting that the situation determines our action. However, we see strong personal differences in other situations. Some people avoid dogs while others go out of their way to pet them. Thus, psychology is confronted with the challenge of accounting for why we respond in distinct ways to what an observer might characterize as the same situation. This challenge is especially daunting when dealing with the behavior of people with psychiatric problems, because they respond in seemingly unfathomable ways to events in their life. And yet Dorothy and

John's behavior started to make sense once I grasped something about their point of view in relation to what was happening around them. As I mentioned in Chapter 1, this is the issue of context. Context refers not to what the situation means to the observer, but what it means to the person who is responding to it. The staff did not realize that Dorothy's reaction was in response to a letter she had received. Nor would a casual reading of her letter have helped. Understanding the significance that the letter had for Dorothy required knowledge of her husband's way of communicating as well as her dependence on him. In the same way, understanding why John became so disturbed when Frank yelled at him required some awareness of how critical it was for John to be accepted by his peers. And there was clearly much more to be understood about John; I regret that I had practically no knowledge of his past. As the philosopher Mary Warnock has rightly noted, "No account of a person could be complete unless it paid attention to how that person has developed through time."[14]

The first principle for understanding disturbed persons is to look for how their behavior and experience are related to situations as they experience them. Of course, this is not any different from how we ordinarily approach our fellow human beings. Nonetheless, instead of looking at how the disturbed person's behavior relates to a context, we often instead dismiss it as "bizarre" (unrelated to anything), explain it in terms of the individual's diagnosis ("he hears voices because he is schizophrenic"), or relate it to some hypothetical neurological or biochemical process in the person's brain. When we resort to any of these devices, there is a real sense that we are no longer listening to or looking at the person as a person.

However, as the above examples have indicated, it is not necessarily so easy to identify the context to which the disturbed person is responding. The key element of the situation may be hidden from the observer. Only Dorothy and I read her letter, and someone not familiar with the situation would have failed to see the significance of what her husband had written. To complicate matters, the meaning of an event may be far from obvious even to the person who is responding to it. A client of mine once described how apprehensive she became when, for the first time in years, she felt genuinely happy. At first, her judgment was that her uneasiness made no sense. It took some exploration before it became apparent that the last time she had been this happy was just before she was told that a close family member had been killed in an accident. An association had been created between feeling happy and terrible things happening. Hence we can see that it is a mistake to leap to the conclusion that there is no relationship between what the person is doing and experiencing and what is happening around him or her just because the connection is not easily established even by the

person involved. For this reason, patience is essential. Julie Sharif, a graduate of the Seattle University master's program in psychology, has written
perceptively about impatience in clinical work: "The impatient therapist is
never impatient on behalf of the client, but only on behalf of the self. It
is the self who is uncomfortable with the present situation, the self who
feels certain goals must be achieved and certain expectations met."[15] We
tend to become uneasy when we do not quickly grasp what is going on for
a client, assuming that this reflects on our limitations rather than the delicacy
and difficulty of the process of coming to an understanding of a life.

A second principle is that even the strangest behavior or symptom is
potentially understandable. A major obstacle to empathizing with the mentally ill is that their actions not only seem out of context or disproportionate
to the situation at hand, but that they seem intrinsically odd or bizarre and
therefore unsettling to those around them. Examples of such actions
abound: refusing to eat to the point of starving oneself, self-mutilation,
responding to imaginary voices, or washing one's hands so frequently that
they start to bleed. Such behavior seems obviously senseless. Yet I would
argue that even apparently self-destructive or bizarre behaviors have a purpose or a function in the person's life, even if this is not at first self-evident
either to the person in question or to observers. This is certainly not an
original insight. Freud insisted that all aspects of human behavior could be
understood if only analysts listened and observed patiently and kept an
open mind. In his classic book *The Psychopathology of Everyday Life*, Freud
provides numerous examples of how various behavioral peculiarities that all
of us are familiar with, such as "slips of the tongue," can be interpreted as
expressions of hidden psychological conflicts.[16] It is not my contention that
Freud was necessarily correct in his view of how symptoms arise, but I do
concur with his emphasis on making sense of behavior.

Again, I turn to stories to illuminate my point. I start with an example
that is all the more striking because it involves an observer who came to an
understanding of a strange symptom in someone who did not communicate
verbally. Barry and Suzie Kaufman started working with children with
autism after their son was diagnosed with this illness. In *A Miracle to Believe
In*, Barry Kaufman writes about how he gained insight into the behavior of
Robertito, a young autistic boy.[17] He was puzzled by the reason why
Robertito continually moved his left arm and leg in a circular, stereotypic
way. Experts in the field of autism would interpret this ritualistic pattern as
a by-product of autism, understood as a neurologically based disorder.
While not necessarily accepting this explanation, Kaufman did not know
how else to understand it. Yet he thought that there was something oddly
familiar about these movements. At a certain moment, he realized that the

boy's movements resembled those that he himself made when one of his limbs fell asleep. This led him to consider the possibility that Robertito was responding to numbness on the left side of his body. But how could he determine if this was true, given that Robertito did not speak? Kaufman gently pricked Robertito with a pin both on his right and left sides; he found that the boy showed a response only when pricked on the right side. Also, Robertito allowed Kaufman to massage his left side even though he would not normally let anyone touch him, and this provided further evidence that his one side was numb. This incident was critical in giving Kaufman a deeper sense of Robertito as an understandable fellow human being and in fostering a relationship between the two of them.

Kaufman's story shows the importance of persistence, receptivity, and imagination. He cared about, and connected personally with, Robertito, and he took seriously his hunch that there was something familiar about the boy's behavior. The insight into the nature of the boy's movement came, as is often the case with solutions to puzzles, when Kaufman least expected it. Kaufman was sitting in a theater and this brought to his mind the sensation associated with the experience of having one's legs falling asleep.

There is a clear lesson implicit in this example. Although Robertito's problems—both the numbness on his left side and his autism—very likely had an organic basis, this in no way makes it less vital that one gain some understanding of his circumstances. People who have a stroke, for example, are suddenly confronted with a world very different from the one they lived in before, whether it is because their movements are now very limited or because they are unable to communicate as they had previously. To assist persons with disabilities in an effective and humane way, one must attempt to get some sense of what they can and cannot do, and how they are responding to their loss of functioning.

The importance of such understanding has been beautifully demonstrated by the neurologist Oliver Sacks who, through his "clinical tales," has allowed us to step into the world of people with neurological impairments. Take, for example, his story about "Witty Ticcy Ray," a man with Tourette disorder.[18] Tourette syndrome is a rare and strange neurological disorder—its key symptoms are what are technically called motor and vocal tics. Motor tics are abrupt involuntary movements, such as twirling or squatting, and vocal tics include making strange noises and, in some cases, uttering obscenities. Ray, a young man of 24, came to Sacks because his symptoms interfered with his daily living, resulting in his losing jobs and creating problems in his marriage. And yet he had successfully incorporated his tics into his life in a number of ways. He was a jazz drummer on the weekends and had found a way to build wonderful improvisation around his tics; similarly, he used his tics to good advantage when he played ping-pong. Sacks pre-scribed Haldol, an antipsychotic medication that has a good record of reducing tics. The result was a disaster. The medication slowed his tics rather than eliminating them and threw Ray completely off balance physi-cally and psychologically. No longer was he a successful improvisational jazz drummer or a ping-pong player whose surprising and erratic shots gave him an edge on his opponents. Lamenting his situation, he told Sacks: "Suppose you could take away the tics . . . what would be left? I consist of tics—there is nothing else."[19]

He had come to be known as Witty Ticci Ray because of his witticism and his remarkably quick movements. Accordingly, Sacks spent three months exploring in-depth with Ray what it might mean for him to live without his symptoms of Tourette, considering the pros and cons and acknowledging the inescapable difficulties that would follow no matter what his decision might be. Ray's decision was to go back on Haldol, and this time the medication worked more effectively. Sacks noted that now that Ray was prepared for the change, the medication was more effective. There was one problem. Even though Ray felt better during the working week, he missed the creativity and virtuosity of his former self. Thus, he decided not to take the medication on the weekends.

What is so admirable in Sacks's approach is that he pays attention to and takes into account the person's entire existence. He fully realizes the importance of addressing the person's experience and values as well as his or her symptoms and problems. Of course, the idea that one should only think psychologically about persons whose problems are psychological in origin makes no sense because it is tantamount to overlooking both the person who is struggling with the illness or disability and the world as he or she experiences it. While Sacks's approach is far from common practice, there are other professionals who show a similar sensitivity to the interrelationship between neurology and experience. Notable among them were the late Donald Cohen and his colleagues.[20]

As I have mentioned, there has been less emphasis on understanding patients as persons within psychology and psychiatry in recent years. Fortunately, however, there are a number of contemporary psychologists and psychiatrists who are humanistically, phenomenologically, or psychoanalyti-cally oriented and who are keenly interested in finding ways of making sense of patients' disturbed behavior. Notable among them are George Atwood, Robert Stolorow, and their colleagues, who have developed an "intersubjec-tive theory" (I will discuss their work more specifically in Chapter 7). Their theory is informed by the writings of Heinz Kohut (see Chapter 2) and other contemporary psychoanalysts as well as by phenomenology. As one would expect from intersubjective theorists, Atwood and Stolorow believe it is critical to recognize that a person's behavior and experience are connected to his or her past and present relationships.

In one article, they use the case of a very disturbed young woman to show how their perspective allows them to understand apparently bizarre symptoms.[21] The young woman had secretly been whipping herself for a number of years, a behavior pattern that she herself found disturbing and puzzling. Through a process of intensive psychotherapy, the meaningfulness and purpose of this pattern started to become evident. When she and her therapist examined instances of her whipping herself, she realized that she did this after she had done something to which her parents had typically responded with anger or physical punishment. For instance, her parents would punish her if she acted assertively, and so, even as an adult, she would feel highly apprehensive after such actions no matter how appropriate they might have been. By whipping herself she took control of the situation rather than passively waiting for something to happen to her. Of course, this is the understanding she arrived at after some time in therapy, while her original experience was just that whipping herself reduced her feeling of being apprehensive and vulnerable. But there was another motive or purpose at work here as well. The situation in which she grew up was so brutal that she dealt with it by dissociating. In other words, she detached herself from her bodily experience. While this process of distancing gave her some sense of protection, it created its own problems insofar as it became a chronic pattern in her life. She became terrified of losing connection with her body and the world of reality, and of being annihilated. Whipping herself, and thus feeling pain, gave her assurance that she actually did exist, that she was not a disembodied spirit.

In her phenomenological study of self-cutting in women who dissociate, Faith Robinson has come up with similar findings.[22] Robinson concluded that self-cutting occurred in the context of emotional chaos, confusion, and strong emotions such as anger and abandonment, but that the specific patterns and meaning of self-cutting depended upon the person's history. In other words, it is important to remember that while apparently irrational self-punishing behavior is meaningful, it is not as if one should assume ahead of time precisely what that meaning is.

One can hardly underestimate the importance of understanding what initially appears to be bizarre or senseless behavior. For patients (or anyone else for that matter), the result is often that one feels less disturbed and less trapped. The woman discussed by Atwood and Stolorow gradually realized that whipping herself was not senseless. This "self-punishment," a practice that she had inadvertently learned from her parents, reduced her apprehension and helped her to feel more alive. Since she arrived at this understanding in collaboration with her therapist, she also had the experience of feeling understood by another person. Feeling understood, as I stated in Chapter 1, helps the person to feel less isolated and more acceptable. In addition, if therapists do not understand the significance of a "symptom," they are more apt to think that it should just be eliminated, and leap to the conclusion that the persistence of the behavior means that the patient is resistant. But insofar as the "symptom" serves a function in the individual's life, it is unlikely to go away unless psychotherapy enables the patient to find another way to address the underlying issue.

The third principle is that knowledge of a person's history helps us to understand his or her actions in the present. This was true in the case of the woman discussed by Atwood and Stolorow, although not so in the case of Robertito. Ordinary life teaches us the importance of history. Think of how often the behavior or outlook of one of your friends, colleagues, or family members started to become intelligible once you heard more about this person's past. The following case shows how listening to a patient's story enabled a therapist to make sense out of symptoms that at first sight seemed completely inexplicable.

The psychiatrist Silvano Arieti discusses the case of Sally in his classic text *Interpretation of Schizophrenia*.[23] Sally's parents explained that their twenty-two-year-old daughter began showing signs of disturbance a few days after her marriage. The couple returned early from their honeymoon because Sally had become very anxious and wanted to go home. When she went to the new apartment set up by her parents for her husband and herself, she was overcome by obsessions, and her movements slowed down to the point that she went into a catatonic state, which is to say that she barely moved. As a result, members of her family had to look after her, even feed and dress her as if she were an infant. She barely spoke, answering questions with just a word or two. Occasionally, though, she was able to move around, especially outside of her home.

Over a number of sessions, Arieti learns about Sally's experience and history. She explained that when she was not in a catatonic state, she had the impression that small pieces or corpuscles were falling down on her body or from her body. She preferred not to move, because she was afraid that any movement would cause the small pieces to fall. Accordingly, she had to reassure herself that small pieces were not falling down, and she had to check herself constantly in an obsessive way. This task was terrific; it kept her in mortal fear of any movement and compelled obsessive thinking from which she could not escape. She used to ask her relatives to help her with searching, to reassure her that no pieces were falling down.[24]

As Arieti continued to work with Sally, she gradually told him what it was like for her to grow up in her family, and slowly the origin of her problems became understandable. Her mother was overprotective to an astonishing degree and made Sally's decisions for her. Not until Sally was in her teens did her mother allow her to cross the street alone, and she did not allow her daughter to go ice-skating or to picnics because these activities were too dangerous! Her mother interfered with all aspects of Sally's life, such as dictating her choice of friends and insisting on being involved in her purchase of clothing, even when she was an adult.

Sally's father made matters worse. He went along with her mother, even though Sally doubted that he was in agreement with her. Accordingly, Sally felt he betrayed her out of weakness. Because she had few friends and because she was not close to either of her siblings, she felt desperately alone and overpowered by her mother. Even at night when she was in bed reading and away from the watchful eyes of her parents, her mother would shout at her to go to sleep.

At eighteen she fell in love with Robert, a young man who took her to New York to see modern art and who introduced her to a freedom that she did not have at home. Her parents thought that art was impractical, while Robert encouraged Sally's long-standing interest in this area. Once the couple started to talk of marriage, her parents put tremendous pressure on Sally to end the relationship and eventually convinced her that it would not work. About a year later she married Ben, a young man whose outlook on life was in conformity with that of her parents. She married Ben in response to her parents' wishes and not because of a heartfelt desire to be with him.

Because the young couple had little money, Sally reluctantly agreed to accept her aunt's spare furniture for their apartment. However, the aunt had a painting that Sally despised, and the parents agreed that it would not enter the new couple's apartment. When they returned early from their honeymoon, Sally found her parents hanging this exact painting over her bed. This was the final straw for Sally. The painting was very traditional; it portrayed French aristocrats in wigs. In other words, it represented the values that her parents had imposed upon her and that she had momentarily been able to push aside when she was involved with Robert. The fact that her parents hung this painting over her bed showed their complete disregard not just of her wishes but of her as a person.

With this knowledge of her background, it becomes possible to see that Sally's behavior, however disturbed, is rooted in an historical context. She grew up to believe that if she made one wrong move (e.g., choosing a friend not meeting her parents' approval), something terrible would happen.

Having her wishes and her autonomy continually undermined, she reached the point where she was afraid of doing anything. Her catatonic state was an expression of her terror of doing the wrong thing as well as her deep resentment toward her family. Through her behavior, she was saying, "You have made me helpless, and have taken everything that I really wanted away from me so now you can take care of me." We can also more readily understand why her honeymoon did not go well.

Arieti reports that Sally recovered completely. She received individual therapy and then she also had a clinical psychologist come in to visit her daily and provide support. After two years of treatment, she was living with her husband (and getting to know him as a separate person rather than treating him as an extension of her parents) and working part-time; she had also found some independence from her parents' wishes.

This case shows how important it is to know something about a person's history. The point is not that Sally's parents "caused" her illness and were to blame for it. Obviously their behavior influenced their daughter, but another person might have responded very differently, defying the parents by running away from home or seeking support from peers. One could also speculate about the background of Sally's mother and wonder what trauma or prolonged difficulty influenced her to become so controlling and apprehensive. There may have been a genetic factor that predisposed Sally to becoming disturbed (this might also be true for Dorothy and John, described earlier), but we do not know whether or not this was the case. Admittedly, it is not typical for someone with a diagnosis of schizophrenia to make a complete recovery, but neither is it unheard of. Recently, Harding, Zubin, and Straus have reviewed a number of long-term studies of the outcome for people who have been hospitalized with a diagnosis of schizophrenia. They found that while the results varied from study to study, most found that at least half of the patients had either a partial or a complete recovery.[25]

There is little doubt that some emotional disturbances are brought about by biological factors rather than life events. For example, an overactive thyroid may produce symptoms that resemble those of agitated depression. It may be that some forms of disturbance included under the broad category of schizophrenia are due primarily to medical factors. However, one should not leap to the conclusion that there are no psychosocial factors involved in a psychiatric disturbance just because such factors do not seem to be obviously present. Contemporary psychology and psychiatry are in a reaction against the apparent tendency in the 1950s and 1960s to assume that "bad mothering" was the cause of schizophrenia and other forms of mental illness. Today, we have a much more sophisticated knowledge of genetic and biological factors that play a role in mental health. Unfortunately, there has not been a corresponding increase in sophistication about studying the intricacies of getting a satisfactory account of a person's history or about clarifying the distinction between identifying the factors that influence the direction of a person's life and looking for someone to blame. If a disturbed person has parents who not only have psychiatric problems but also act in ways that are disturbing to their son or daughter, then one has to take this into account. However, it does not follow that the parents could simply have chosen not to be disturbed any more than the patient could have chosen to be free of mental illness. In other words, it makes no sense to assume that the concern with looking at the historical and interpersonal context for a patient's problems inherently amounts to an accusation against the patient's parents or caregivers, or that it is inherently a process of seeking to place blame.

Getting adequate background information on a patient is no easy matter, and in many cases it simply does not happen. These two points were documented by a study carried out by Melitta Leff, John Roatch, and William Bunney in 1970.[26] They were skeptical about the findings of several earlier studies that suggested that many episodes of depressive disorder developed in the absence of precipitating factors. These studies relied upon information collected early in the hospitalization when the patients were the most disturbed and confused. To correct this limitation, Leff and her colleagues studied forty hospitalized patients over a number of months. Most of them had previously been depressed, many of them had attempted suicide, and they had classical symptoms of depression such as feelings of worthlessness and despair, sleep disturbance, and change in weight.

Leff, Roatch, and Bunney were thorough in investigating the patients' past: they got their data from twice-weekly therapy sessions conducted for as long as the patient was in the hospital, weekly interviews of spouses or family members of the patients, and information and observations from ward staff. Using this approach, they found that, on average, each patient had four significant stressful events that occurred preceding the onset of the disturbance. The most frequent types of events included threat to sexual identity (e.g., being told that one is infertile), changes in marital relationship (e.g., one's spouse threatens divorce), a geographical move (and thus loss of connection to friends and neighbors), and compulsion to face a denied reality (e.g., that one's father was really dead). These are major events and yet few of them were mentioned early in the hospitalization either by the patient or by his or her family. Only in two patients out of the forty did they find a pattern of mood swings that seemed unrelated to external events.

It should not surprise us that it takes a special effort to find out about a person's history. Our own experience with people we know intimately tells us as much. How much do we know about our parents' past, our friends' deepest fears, or our siblings' love relationships? We may know very little about our parents' past because we lack interest or do not think of asking them questions. There is also a great deal that parents prefer not to tell their children, perhaps with the conscious intention of protecting them or because of shame and embarrassment. Sometimes parents tell their children so many stories that it never occurs to the children that just as many stories remain untold.

If we keep so much of our own history hidden even from those who are close to us, we should not be surprised that people with serious psychological problems conceal details about their past from their therapists. The psychiatrist Harry Stack Sullivan has commented that those who interview patients often fail to arrive at a meaningful understanding of them. Sullivan realized that learning about another person is a very delicate task, requiring a great deal of skill and, above all, a trusting relationship with the patient.[27] Much of what one initially hears from a patient may turn out subsequently to be misleading or limited in scope. We guard the story of our lives until such a time that we feel that we can tell it in safety.

What I am suggesting is that the process of coming to an understanding of a person who is mentally ill is fundamentally the same as that of coming to understand anyone else. Because people who are seriously disturbed may be particularly unable or unwilling to tell us what is going on for them, the process is more likely to be difficult. And this is also apt to be true because their experiences may be beyond what any of us are readily able to imagine. Nevertheless, the process of reaching a deeper understanding of another person is fundamentally the same whether the person is well or "ill." In arguing that people who are disturbed are more like than unlike us, I am simply echoing what psychotherapists who have worked intimately with patients diagnosed with schizophrenia, such as Sullivan, Frieda Fromm-Reichman, Ronald D. Laing, Bertram Karon, Gary Vanden Bos, and Garry Prouty, have tried to tell us.[28] As Fromm-Reichman has written, if one regards patients as "strange creatures of another world" whose behavior and speech are not understandable to normal people, then one cannot treat them. Genuinely meeting one's patients requires that one realize that the difference between oneself and them is "only one of degree and not of kind."[29]

To argue for the possibility of some degree of understanding of, and even meaningful contact with, people with serious mental illnesses is by no means to minimize the seriousness of their disturbance. Anyone who has worked or lived with someone with a psychiatric diagnosis of schizophrenia, bipolar disorder, or borderline personality disorder, for example, knows all too well how devastating such disturbances can be, both for the patient and for those who care about him or her. Rather, to affirm the possibility of understanding is to remind us of a possibility that has already been documented time and time again by psychotherapists and lay persons alike and to take seriously what a number of patients themselves have said about how contact with another can aid in healing. One recovering patient has written eloquently to that effect:

> The question of whether the fragile ego of the schizophrenic patient can withstand the rigors of intensive therapy seems to me an unfortunate hindrance to the willingness of psychiatrists to attempt psychotherapy with schizophrenic individuals. A fragile ego left alone remains fragile. It seems that there must be some balance that can be achieved so that schizophrenic patients can receive the benefits of psychotherapy with therapists who are sensitive to their special needs and can help their egos emerge, little by little. Medication or superficial support alone is not a substitute for the feeling that one is understood by another human being. For me the greatest gift came the day I realized that my therapist really had stood by me for years and that he would continue to stand by me and to help me achieve what *I* wanted to achieve.[30]

Depersonalizing Trends in Mental Health

Having referred to the current depersonalizing trends in psychiatry and psychology in passing, I want to say more about them here. However, I want to make it clear that I am not writing a polemic against biological-oriented approaches to psychiatry or a manifesto against the use of medication in the treatment of mental illness. Judicious use of medication has an important place in psychiatry, and innovations in the study of genetics and neuroscience contribute to a better understanding of the physical underpinning of psychiatric disturbances. These perspectives illuminate only a limited aspect of the whole person, and yet they have often claimed far too much territory for themselves with obviously troubling consequences. By necessity my critique will be brief, but in the text and in endnotes I refer to additional authors and material that address these issues in greater depth.

Let me start with a scenario, familiar to most psychotherapists, that illustrates some of these troubling consequences. Sharon, a woman who had started therapy with me because of ongoing depression, told me during one of our sessions that her problem was caused by a deficit of serotonin (a neurotransmitter) in her brain. Here she was repeating what her physician had told her. This "serotonin theory" has been treated in the popular press and among some professionals as established fact. How did her physician know the state of her brain? The fact is that he simply assumed that this theory was correct and applied it to her.

I did not directly respond to her comment, but over the next several sessions it became apparent that her depression was again "sneaking up" on her. One setback was discouraging to her, but she put it aside. Something else happened that was unfortunate, increasing her discouragement. Then a third incident came along, adding to her existing self-doubts and distress, and eventually she became overtly depressed. The "serotonin theory," presented as a comprehensive explanation of depression, was a distraction. Fortunately, Sharon did not let this explanation deter her from exploring aspects of her life that rendered her vulnerable to becoming depressed. However, theories of this sort, if taken too literally or as providing the "complete truth," can take us away from listening to ourselves and to others. The risk is that one substitutes an abstract explanation for a careful examination of one's actual circumstances and how one responds to them.

Anyone who looks at a current abnormal psychology textbook will find that this rather simplistic theory has been abandoned. Whatever is going on in the brain when people get depressed is more complicated than a drop in the level of serotonin, and whatever antidepressants do is more elusive than causing an increase in serotonin.[31] Moreover, it would be a mistake to assume that a decrease in serotonin (or whatever else is going on in our brains and bodies when we are depressed) is the cause of depression. According to the Leff, Roatch, and Bunney study, depression follows particularly difficult events in people's lives. This suggests that biological changes do not typically cause but instead accompany depression.

As I mentioned earlier, innovations in psychiatry have paradoxically taken some of our attention away from patients as persons. The surgeon general's report enthusiastically comments that "today, integrative neuroscience and molecular genetics present some of the most exciting basic research opportunities in medical science," and Nancy Andreasen, a well-known psychiatric researcher, points out that "brain research on mental illness has made substantial advances in recent years supported by conceptual and technological developments in cognitive neuroscience."[32] There is nothing inherently wrong with this enthusiasm, and neither source commits the error of asserting that mental illness can be understood just by studying the brain. The surgeon general's report emphasizes the importance of the environment both in the treatment and in the development of psychiatric disturbances, and Andreasen is careful to point out that while the brain affects experience, experience also affects the brain. Nonetheless, as the psychiatrist Elio Frattaroli has pointed out in his recent book, we are living in an age where preoccupation with the brain has led to a neglect of the soul and where a very narrow definition of science as laboratory based is predominant. Instead, he argues, as did Amedeo Giorgi thirty years earlier, "The first rule of science is that *the method of observation must be appropriate to the phenomenon being observed*" [italics in original].[33] To study psychiatric disturbances, in other words, you have to study people's experience and circumstances, and not just their brains.

However, these views—that the biological is but one aspect of the person and that more holistic approaches are called for—tend to be forgotten or are merely given lip service in professional circles and in the media. Discussions of psychiatric disturbances in the popular media are increasingly based on the assumption that they are "brain disorders." One of the factors that leads to such simplifications is that most reporters or other journalists have little understanding of psychological or biomedical issues. But this is by no means the only factor. The belief that severe mental illnesses are known to be physical brain disorders has been actively promoted by several groups and for various reasons. The most powerful mental health advocacy group in the United States, the National Alliance for the Mentally Ill (NAMI), claims on its web page (http://www.nami.org) that this is true not only of schizophrenia and affective disorders but also obsessive-compulsive and anxiety disorders. NAMI was founded by parents of people with mental illness who wanted to promote good treatment for the mentally ill. Such advocacy is commendable and much needed. However, this concern by itself does not explain or justify the insistence on a one-dimensional understanding of psychiatric disorders. Even a cursory examination of the newest diagnostic manual published by the APA, an organization that by no stretch of the imagination can be described as downplaying the importance of biomedical factors in mental illness, would lead one to wonder why NAMI insists that these are "physical brain disorders." For example, the section on schizophrenia in the manual states that "there are no laboratory findings that are diagnostic of schizophrenia" (that is, a person is diagnosed as schizophrenic on the basis of behavior and reported experience) and that while genetic factors play an important role in this disturbance, so do environmental factors.[34] Of course, it is possible that at some point in the future our understanding of schizophrenia will move toward the position for which NAMI currently advocates, but why would one leap ahead of the evidence? Furthermore, why would one also include anxiety and obsessive-compulsive disorders where the evidence for the interpretation that they are "physical brain disorders" is much weaker?

Part of the answer has to do with the concern about overcoming prejudice toward the mentally ill. As Nathaniel Lachenmeyer, who has written a powerful book about his father's psychiatric disturbance, has rightly said, "People with schizophrenia not only face their symptoms; they face pervasive prejudice."[35] Even though Lachenmeyer appears to subscribe to the theory that schizophrenia is a neurobiological disorder, one of the major contributions of his book is that he makes sense out of his father's life. He attempts to discern his father's intentions, even during periods when he became increasingly delusional, and takes into account significant events, such as the failure to get tenure and the death of his mother, that contributed to his father's downward spiral. It is hard to imagine that one could reasonably interpret the life of a person with Alzheimer's (clearly a physical brain disorder) in similar fashion. Insofar as this book contributes to overcoming prejudice toward those with schizophrenia, I believe it does so largely through showing the basic humanity of someone who was very disturbed.

Nonetheless, the primary tactic that has been used in campaigns to reduce prejudice against the mentally ill has been to insist that mental illness is like every other illness, in other words, that it is a "physical brain disorder." From a public relations perspective it makes sense strategically insofar as the general public and even some professionals tend to think in terms of "either/or." That is, many people appear to believe that there are just two possibilities: either mental illness is a "real" illness that can be treated medically, or it is just in the person's mind. If it is not a medical illness, then it is just a mental or a psychological problem, and a person should be able to snap out of it. Psychological or psychiatric problems, following this line of reasoning, are viewed as "weaknesses," something that one is protected against if one lives an upright life or strives to maintain a positive outlook. Alternatively, if the disorder is thought to be related to the environment, parents are to blame. This is another reason to insist that mental illnesses really are "physical brain disorders." Yet the obvious problem with an exclusively biomedical interpretation is that it overlooks the complexity of the evidence available to clinicians and researchers.[36] And, as I have emphasized throughout this chapter, it gets in the way of our listening to the stories that people with diagnoses have to tell.

Of course, another factor in giving emphasis to a medical interpretation is economic. The pharmaceutical companies that manufacture psychiatric medication have enormous advertising budgets. In 1996 they spent $600 million on advertising aimed directly at consumers, ten times as much as they spent in 1991 for the same purpose.[37] The lobbying group for these companies, the Pharmaceutical Research and Manufacturers of America (PRAMA), spent about $150 million in 2004, lobbying Congress and state legislatures and giving support to organizations and economists that favor their point of view;[38] in 2002 PRAMA gave roughly $27 million to political candidates.[39] This is an enormous amount of money that enables this industry to have significant influence on the legislative process. Moreover, as the journalist Cynthia Crossen has pointed out, "The pharmaceutical companies are master marketers."[40] They have a vested interest in selling their products and therefore they also have a vested interest in persuading people who are depressed, anxious, socially awkward, and so on, that their problems have a medical basis and that a medical remedy is not only appropriate but the preferred mode of treatment.

In any case, this industry has made huge profits from their products. In the year 2000, for example, Americans spent $10.8 billion on antidepressants, a 21 percent increase over the previous year.[41] Eli Lilly, the manufacturer of Prozac, relied heavily on the profits from this best-selling antidepressant for a number of years. It is estimated that it was used by about forty million people in ninety countries in 2001.[42] The number of American children receiving psychiatric medication tripled from 1987 to 1996 and continues to rise.[43]

None of this is to deny that medication can be helpful, or even essential, in specific cases (e.g., for those with bipolar disorders). Some people who take antidepressants report that they seem to work miracles, at least at first. Sharon, the woman who saw me in therapy, felt that her medication was moderately helpful. But medications have side effects (though often fewer than in the past), they sometimes make things worse, and, even when they are effective, they by no means address all of a person's problems. Just ask the spouse or close friend of someone who takes antidepressants. David Karp, who interviewed fifty people with depression for his book *Speaking of Sadness,* found that most of them became "disenchanted with the value of medication for solving their problems" even though many of them continued to take it.[44] Karp, a sociologist, is one of the very few social scientists who has taken the time to find out what it is like to be depressed and how patients experience psychiatric treatment.

There appears to be less and less emphasis on patients' experience in psychiatry. This is evident when one looks at the changes that have been made in the diagnostic manuals developed by the APA, starting in 1980. The third edition of the manual, published in that year, is dramatically different from its predecessor, published in 1968. The earlier two versions had brief and often vague definitions of disorders. The 1980 edition and the subsequent editions and revisions (1994, 1997, and 2000) have far more sharply defined criteria for each mental illness, are based on extensive research and clinical trials, and are written by "work groups" of psychiatrists and other mental health practitioners. Prior to 1980 the practice of psychiatric diagnosis was much maligned. Psychiatrists did not easily agree on a diagnosis because the criteria they worked with were largely imprecise. Critics of the practice of applying "labels" to people with a mental illness diagnosis could point to study after study that suggested that the whole process was unreliable and unscientific and did not lead to effective and humane treatment.

However, starting with the 1980 revision of the manual, the DSM–III, the phoenix of diagnosis, rose from the ashes. The psychiatrist Gerald Klerman was undoubtedly right when he wrote that the "DSM-III has already been declared a victory. There is not a textbook of psychology or psychiatry that does not use the DSM-III as the organizing principle for its table of content and for classification of psychopathology."[45] Based on careful research, and with precise definitions and explicitly spelled-out criteria, the new generation of manuals allowed clinicians to reach consensus on diagnosis to a much greater extent than previously. These manuals have been accepted as authoritative within the mental health community and have also provided the APA with considerable income from the sale of well over half a million copies of each version.

From a humanistic perspective, the newer manuals do have some positive features. They avoid equating a person with his or her diagnosis, substituting terms such as "a person with schizophrenia" for "schizophrenic," and they acknowledge that behavior is always situated in a cultural context. On the other hand, these manuals have far more credibility than the pre-1980 versions, with the result that diagnostic categories are taken much more seriously and more literally by both mental health professionals and patients. Why is this a problem?

First, the whole practice of psychiatric diagnosis is based upon on a medical model. Yet it is not self-evident that psychiatric problems fit this model. The manuals define "mental disorders" in quasi-medical terminology while also acknowledging that this term is ambiguous.[46] One problem with the immense success of the new diagnostic manuals is that it conceals the questions regarding the nature of psychiatric diagnosis rather than addressing them. One of the critics of the manuals, George Vaillant, has argued that psychiatry is more of a poetic science than a hard science, and therefore the relatively clear-cut distinctions that characterize current diagnoses are inappropriate. Others have echoed his comments, noting that much of what psychiatry deals with are human problems, not medical illnesses.[47]

Secondly, there is less emphasis on the patient as a person. Ellen Corin and Gilles Lauzon, psychologists who have specialized in the study of persons diagnosed as schizophrenic, argue that with the new manuals there has been a shift in perspective, a shift that has resulted in "the bracketing-off of any subjective dimension and the correlative exclusion of meaning from the realm of scientific knowledge and practice."[48] Similarly, Lawrence Hartman, a former president of the APA, has stated: "They [the manuals] emphasize clarity and validity but, many clinicians think, sacrifice validity and the whole person."[49] As Philip Cushman has pointed out, "The DSM-IV self is a self of parts, not wholes; behaviors, not personalities; concrete observations, not artistic interpretations."[50] The new manuals are certainly very structured, bureaucratic documents that treat diagnosing as if it were something akin to an exact science. Essentially, the criteria are presented in the form of checklists, where one has to have two out of four symptoms from category A, three out of five from category B, at least one from category C, and so on, to "qualify" for a diagnosis. Yet, in actual practice, "clinicians often claim that the more they know patients, the greater difficulty they have in fitting them into a category."[51] Increasingly, though, published case studies have given a great deal of attention to how particular patients fit within categories and much less attention to the particular circumstances and histories of these patients. To put it simply: diagnoses overshadow persons.

One might think that psychologists, given that they are not typically committed to the medical model, would give some attention to the study of the experience of people with psychiatric disturbance. Unfortunately, as I indicated in Chapter 3, psychologists rarely use qualitative data. When Karp became interested in depression, he reviewed the professional literature and found that it included the views and observations of all kinds of experts but "never the voices of depressed people themselves."[52] This pattern is true not just of literature in depression but of other areas as well. For example, Whitaker, in his disturbing book *Mad in America,* has noted that there that are virtually no references in the psychiatric literature to how patients with schizophrenia experience or think about the medication they take.[53] Disregard of patients' experience is hardly a sign of progress.

To an extent that is not often acknowledged, even psychologists are living in the shadow of the biomedical model. As we have seen, the American Psychiatric Association has done an impressive job of persuading professionals, insurance companies, and the general public that diagnosis is a credible and necessary psychiatric practice and that its recent manuals are exemplary. There are certainly critics of current diagnostic practice, and a good number of them are psychologists, but they are in the minority. Every single abnormal psychology textbook that I have seen in the last twenty-five years has been organized around the new manuals. This is but one of many indicators of the extent to which the psychology establishment has embraced this new (post-1980) approach to diagnosis. Psychologists are also keenly aware that psychiatric medication has come to play a major role in treatment. Although many psychologists have a healthy skepticism about the claims made for the effectiveness of such treatment (especially if used apart from other interventions), some have called for prescription privileges for psychologists with special training in psychopharmacology and related areas. This move has been opposed, not surprisingly, by psychiatrists, and is one of the most hotly debated topics within the profession itself.

Yet there is no question that psychologists have contributed a great deal to the field of mental health with respect to research, assessment, and treatment. There is plenty of evidence that psychological treatment is effective with a variety of psychiatric disturbances, ranging from depression to anxiety disorders. Nonetheless, a number of psychologists are quite concerned that therapy will be left in the dust by psychopharmacological treatments unless there is a concerted effort to "officially" identify specific psychological treatments that have a record of success with particular disorders.[54] In 1995, a task force within the American Psychological Association issued an initial report giving examples of what they called "validated" treatments (e.g., cognitive therapy for depression), meaning that the task force believed that there was research that demonstrated these treatments had a good track record. This may seem like a rather technical issue that is far removed from the basic concerns that I am addressing in this chapter. However, the connection is clear once we consider that all of the treatments the task force "validated" were ones that used treatment manuals. For example, there are manuals specifying how to treat panic disorder or depression following a cognitive model. Having therapists follow a manual presumably increases the extent to which the therapists using a given approach treat patients in a similar fashion. Accordingly, one has greater confidence that any success is largely attributable to the techniques used rather than to the skills, personality, and wisdom of particular therapists. One "pro-manual" psychologist wrote, "If psychotherapy is only an art, why do we need doctoral programs? I do agree that there is an art to psychotherapy as well as science, but my own goal is to steadily increase the science, not to entirely trust in art."[55]

This is a depersonalizing move, an attempt to minimize the personal aspect of the psychotherapy process. Moreover, this approach also depersonalizes the patient or client because, by definition, manuals are written for the treatment of specific problems, such as panic attacks, and not for the treatment of particular people. In practice, however, clients do not bring in just one particular problem, and even if one could conceive of such a thing, it is obvious that the very nature of the problem would take on its specific shape and color according to whose problem it was and what his or her life circumstances were. A number of psychologists have rightly complained that the philosophy underlying this whole approach is highly mechanistic, and that the emphasis on the use of manuals excludes psychotherapy based on humanistic and relational principles.[56] Not surprisingly, some clients refuse to participate in psychotherapy where the therapist follows a manual.

To some extent the call for validated treatment was based on the concern that psychological therapies were losing out to psychiatric medication even though the evidence for the effectiveness of the former was strong. But if the evidence was already strong, why the push to arrive at a list of validated treatments? The psychologists involved wanted to come up with a relatively precise specification of what treatments work for what disorders and, further, to identify the specific dimensions of each treatment that account for its success. This is exactly what one does in research on medications. Antidepressant X is compared with a placebo and perhaps with another antidepressant in order to establish that the medication is better than an inert substance (the placebo) and to further determine which medication is more effective. Then there is further research to determine what effects the medication has and which of these effects are critical in reducing depression (the other effects, the ones that one does not want, such as constipation and nausea, are called side effects). One of the goals of using manuals is to minimize the effect individual therapists have on the treatment. In other words, the more precise the manuals are, the less the need for the therapists to use their own judgment as to how they should respond at a given point. In addition, the more precise the manuals, the better the chance that one can identify the particular critical components in the treatment that bring results. Essentially, this is an attempt to factor out the human dimension of psychotherapy. Thus, eventually, one would be able to say with authority that therapy X is more effective than therapy Y for depression, for specific reasons, just as one supposedly is able to do with different kinds of medication. Therefore, it is not just that psychologists live in the shadow of the medical model, but that some psychologists actively embrace it.

In embracing this model, the philosopher William Barrett would say, psychologists (and many others) are falling prey to illusion. In his 1979 book *The Illusion of Technique,* Barrett has suggested, as I mentioned in the previous chapter, that a technique is a standard method that can be taught because its steps can be precisely specified.[57] He further suggests that belief in techniques as a way to deal with human problems becomes superstition insofar as the transposition involves three assumptions that he believes are unsupportable. First, techniques exist in and of themselves. That is, a technique can be understood and used apart from any particular cultural or personal context. In this way, a psychological technique would be effective (if one follows the manual correctly) no matter who uses it or in what context it is used. Second, techniques can be applied to human affairs in a way that is very similar to how they are used in dealing with the natural world. Third, techniques can provide basic solutions to human problems; that is, they allow us to escape issues such as ambiguity, relationships, and interpretation.

Ursala Franklin, an experimental physicist and a critic of the role of technology in modern society, distinguishes between prescriptive technology, which is what Barrett addresses, and holistic technology, which involves discretionary judgment, creativity, and attention to human relations (the very thing that therapy manuals are trying to eliminate as much as possible).[58] I seriously doubt that any psychologist really believes that it is possible or desirable to turn psychotherapy into a set of techniques that can be spelled out in a manual. However, the degree to which faith in techniques and technology is professed in our society is certainly remarkable. In this regard, the historian David Noble has suggested that in the Western world, and especially in North America, one can appropriately speak of the religion of technology.[59] Techniques have been so effective in the area of mass production that it is tempting to think it can also be applied to all of those daunting human problems that we are reluctant to tackle on a personal level.

My colleague Erica Lilleleht, in examining the field of psychiatric rehabilitation, has discovered how emphasis on techniques predominates even in an arena that espouses humanistic values. The particular tradition that she has focused upon was developed by William Anthony and his colleagues at Boston University's Center for Psychiatric Rehabilitation.[60] According to their foundation texts, their program is structured to help psychiatric patients return to living in the community and emphasizes the teaching of skills, questions the usefulness of diagnostic labels, and extols the importance of seeing the clients as persons and of supporting them in reaching the goals that they set for themselves. However, the manuals that are written for the professionals and paraprofessionals who will work with these clients embody a very different philosophy of treatment and of the person. As Lilleleht points out, "These manuals pursue skill development using a rhetorical style and a set of teaching practices that are highly *reductive,* with a strong emphasis on *division* for the sake of observation and measurement" [italics in original]. In contrast to the holistic emphasis of the foundational texts, these "training manuals teach participants to identify, divide, subdivide, and evaluate every component of any relevant skill."[61]

Here we find ourselves back with Barrett's definition of technique as involving steps that are precisely specified, and his assertion that the belief that such techniques work in the realm of human relations amounts to a superstition. In the previous chapter on forgiveness, we saw that psychologists have developed procedures for helping people move toward forgiveness. My own judgment is that many psychologists move back and forth between the idea of prescriptive technology (identifying minute steps that can be followed precisely) and holistic technology (following general principles and using discretionary judgment), but that prescriptive technology has a very seductive appeal because of its promise of control over human behavior, as in the above case of rehabilitation psychiatry.

Conclusion

The surgeon general's 1999 report on mental health laments the stigma that the mentally ill face but expresses the hope that high-quality research will overcome people's misconceptions about mental illness.[62] It is not clear exactly what kind of research would accomplish this goal, but presumably this report is a reference to studies on the genetic and neurological aspects of mental illness. Also included might be more sociologically oriented research that would deflate the myth that people with psychiatric problems are more dangerous to other people than the rest of us and give us in its place a better understanding of the difficulties that these people encounter in their daily lives. Nowadays, psychological and psychiatric issues are discussed more openly than was the case in the past, and it is no doubt fair to say that research has contributed toward the goal of overcoming stigma.

But, as I have noted, the research on the biological and genetic aspects of mental illness has created quite another problem, one that I have referred to throughout this chapter. This is the problem of the objectification of self and others and what I have called the disappearance of the mentally ill. Constance Fischer, whose work on assessment is discussed in Chapter 2, writes about the clients who come to her office and speak of themselves in objectifying, totalizing terms. They explain their problems in terms of their diagnosis or other psychological constructs such as inferiority complexes or intelligence quotients.[63] With respect to objectification of the other, Frattaroli points out that "by thinking of the patient as having a brain defect that makes him *different* [italics in original], the doctor avoids the anxiety of having to feel his [sic] common humanity with his patient, and having to recognize that the patient's illness reflects an existential dilemma to which the doctor is vulnerable as well."[64] The point here is not to deny differences or distinctions between patient and doctor, between the person who is disturbed and the person who is not, or between the person who has a neurological problem and the person who does not. Rather, the point is to assert that we share a common humanity, notwithstanding these differences, and that this common humanity provides for the possibility of a genuine relationship and at least some degree of understanding. As the patient with the diagnosis of schizophrenia wrote about the value of psychotherapy (given earlier) stated, "Medication or superficial support alone is not a substitute for the feeling that one is understood by another human being." The challenge, of course, is developing a relationship. Coming to understand another person is *not* a technical enterprise. On the contrary, we have to become present to another, open ourselves up to and learn from him or her, and also learn about ourselves if we truly want to understand someone. This is a challenging process under the best of circumstances, and it is especially challenging (as well as rewarding) if the other person is seriously disturbed. As I indicated at the beginning of this chapter, recognizing the humanity of someone who is different from oneself in some way deepens our sense of our own humanity. It is a gift that benefits both the recipient and the person who gives.

Finally, I want to return to the issue of how the biomedical model has implications for everyone—the study of brain chemistry, neurology, and genetics are relevant for an understanding of all human beings. We should be aware that there is not a clear-cut, medically based dividing line between the normal and the abnormal. As Andreasen writes, "These are boundaries of convenience that permit reliable definition, not boundaries with any inherent biological meaning."[65] The distinction is based primarily on psychological and social criteria, and these are by no means as precise as some defenders of the new diagnostic manuals would have us believe. Moreover, as Elliot S. Valenstein has pointed out in his examination of the relationship between drugs and mental health, "Contrary to what is often claimed, no biochemical, anatomical, or functional signs have been found that reliably distinguish the brain of mental patients."[66]

Of course, cognitive neuroscience—the approach for which Andreasen advocates so effectively—studies all kinds of neural mechanisms underlying cognitive processes, be they regarded as normal or abnormal. One could as well study the brain or mind of the chess player or the person in love as that of the person who is diagnosed with depression or the one who has hallucinations. This is powerful stuff—it is all about science, sophisticated technology, and the promise of application. But it also provides a very narrow lens through which to look at human beings. William Bevan, in his critique of contemporary psychology, has argued that the notion of the brain/mind as a "versatile computer" is both reductionistic and mechanistic.[67] In addition, cognitive neuroscience looks at the person in isolation from others and outside of a social context. Admittedly, there is the acknowledgment that the environment plays a role. Yet technological developments do little to deepen our understanding of the interpersonal, the historical, and the social. Moreover, cognitive neuroscience has little to say about these matters because you cannot see or relate to a person through computerized tomography and magnetic resonance imaging.

The previous chapters, as well as this one, are based on stories about getting to know or looking anew at another person and thereby also at oneself. These experiences—of seeing another as if for the first time, being disillusioned, forgiving another, and developing empathy for a person who is disturbed—shatter any illusion we might have that human existence can be understood in biomedical terms. Nonetheless, illusions about the power of science persist. Recently, I heard a high-ranking scientist/administrator at the National Institute of Health proclaim during an interview that the brain is the only organ that is capable of reflecting on itself. This is nonsense, of course. The brain does not think or reflect, the person does. We have to be careful about what hopes we place on science and technology for giving us an understanding of ourselves and for addressing human problems.

I doubt that anyone believes that bias toward the mentally ill will ever be eliminated, however much it is a goal for which we ought to strive. Nor do I believe that one should rely just on scientific studies to reduce such bias. People's stories we read or hear about are at least as important in bringing us into closer contact with those psychiatric disturbances. Lachenmeyer's moving account of his effort to understand his father's disturbance is a good example of such a story. It is a book that has the power to move readers beyond their preconceptions of the mentally ill, without in any way romanticizing or minimizing the agony and tragedy with which patients and their families often have to deal. Movies based on books, such as *I Never promised You a Rose Garden* and *A Beautiful Mind* (both true stories), also play an important role.[68] But, above all, we are influenced by our own relationships with those with psychiatric disturbances as well as whatever emotional problems we ourselves struggle with at various points in our lives. To help us with all of this, we have to call upon psychiatry and psychology as "poetic sciences," to use Vaillant's expression, rather than just as biomedical sciences. We need approaches that reflect back to us the fullness of our humanity. Among these approaches we must include the humanities, including the arts. This is an issue that I will explore in the next chapter.

CHAPTER 5

On the Study of Human Experience

The ultimate test of a study's worth is that the findings ring true to people and let them see things in a new way

David A. Karp[1]

When we psychologists, perennially anxious about our identity as scientists, insist that science making is totally different in kind from other exercises of the human intellect, we are perpetuating a false dogma, the effect of which on the future of the field can only be pernicious.

William Bevan[2]

What is phenomenology and how does one practice it? In this chapter I will attempt to answer these two questions in a non-technical way and will also show that phenomenology is not esoteric but an approach that intuitively makes sense. However, by stating that this approach makes sense intuitively, I do not mean that phenomenology is in accord with common sense, understood as the array of opinions that a group or an individual takes for granted as being self-evidently true, that is, the clichés, prejudices, or assumptions that all of us carry around with us. Rather, by speaking of the intuitive, I am referring to insights or understandings that relate to our immediate experience of the world and of ourselves. Mainstream psychology textbooks also point to how research findings contradict common sense, but imply that our experience of the world is unreliable. In their classic social psychology textbook (1967), Edward Jones and Harold Gerard cite criticisms directed at a large-scale study conducted by Samuel Stouffer and his colleagues of the professional and psychological adjustment of American soldiers during World War II.[3] The critics basically argued that the conclusions belabored the obvious. Jones and Gerard also quote at length the spirited defense of such studies by the sociologist Paul Lazarsfeld, who pointed out that many of the

conclusions of Stouffer and his colleagues were contrary to what most people would have expected.[4] For example, the researchers found that black
soldiers were more highly motivated than white soldiers to become officers
and that city boys had higher morale than farm boys during basic training.
This same study is discussed, in almost exactly the same way, in Michael
Passer and Ronald Smith's recent introductory psychology textbook,
although without any mention of Lazarsfeld's response.[5]

Although these findings might well have been surprising to many
people, one would be hasty in jumping to the conclusion that studies of
this sort demonstrate the inherent limitations of people's observations.
A more reasonable conclusion is that one should not trust the opinions
(however widely shared) of those that have little or no direct experience
with the issues at hand. The more pertinent question is whether these
findings were surprising to those with extensive experience in the armed
forces during this period. In a review of introductory psychology textbooks, Robert Romanyshyn and Bryan Whalen found that many of their
authors argued that true or scientific knowledge requires that one leave
behind direct observations and arrive at some fundamental reality that
lies behind them, using experimental and other specialized methods.[6] In
contrast, the goal of phenomenology is to be faithful to experience, but
this does not imply an endorsement of common sense understood as
preconceived opinions.

In the Introduction and the first four chapters of this book, I have
already described phenomenology (and phenomenological psychology, in
particular) as an approach to studying human life that regards stories and
other first-hand accounts as valid sources of data. Phenomenology
expresses its insights in language that does justice to human experience
in order to deepen our understanding of our existence. In Chapter 2, on
disillusionment, I discussed Constance Fischer's distinction between primary and secondary data. The former refers to what we experience and
observe directly, and the latter refers to the interpretations and conclusions that we, as ordinary persons and, especially, as social scientists,
draw from the primary data. For example, we notice that one of our
friends minimizes his achievements and compares himself unfavorably
with other people (primary data). To account for this puzzling pattern
of behavior, we fall back upon the ready-made explanation that this man
has an "inferiority complex." By embracing this explanation, we step back
from our direct experience of our friend and effectively close off further
exploration of what this behavior means for him and for his relationship
with others. That is, his behavior is now explained with reference to some

hypothetical mechanism (a complex) inside him that causes him to act in a certain way.

The Core of Phenomenology

None of this is to suggest that phenomenologists avoid conclusions or generalizations about particular experiences or phenomena. The preceding chapters have included generalizations about epiphanies in relationships, disillusionment, and forgiveness, as well as stories and descriptions. However, phenomenology (and existentialism, with which it converged by the middle of the twentieth century) seeks to take us back to phenomena so that we might understand them more fully, not just as individual experiences but as having common themes or qualities. If we listen to a number of stories about forgiveness, for instance, we soon start to realize that they have something fundamental in common, such as shedding resentment, finding healing for oneself, and seeing the wrongdoer as a fellow human being. To acknowledge these dimensions of the experience does not mean that one substitutes a theory for a phenomenon, or abstraction for concrete experience. Instead, such understanding can allow for a deeper appreciation of a phenomenon, providing that one does not treat the general description as more real than the experience itself or as a final truth about its essence. There is always more to be said, different perspectives from which to say it, and a variety of ways of saying it.

Karp's statement about what makes a study valuable (the chapter epigraph) succinctly describes the essence of phenomenology. His statement also affirms that phenomenology is an approach that intuitively makes sense. In the previous chapters I have presented findings on specific phenomena and, as reader, you have had the opportunity to evaluate whether, in light of your own experience, the findings ring true and helped you to see something familiar in a fresh way. In everyday life each one of us is something of a phenomenologist insofar as we genuinely listen to the stories that people tell us and insofar as we pay attention to, and reflect on, our own perceptions. This means that phenomenology is practiced, at least informally, by people who have never heard the word or know little or nothing about it as an intellectual tradition. Maurice Merleau-Ponty suggested as much when he asserted that "phenomenology can be practiced and identified as a style or manner of thinking, that it existed as a movement before arriving at complete awareness of itself as a philosophy."[7]

Of course, some of us are more attentive to, and more thoughtful about, what we observe than are others. Some people are such good listeners that they quickly grasp the essence of what is being said. Artists are often especially adept at portraying or speaking to something fundamental in human life. Consider the following lines, from the Canadian poet Miriam Waddington:

> There is a man who calls me wife
> who knows me but does not know my life
> and my two sons who call me mother
> see me not as any other
> yet if the fabric of my day
> should be unwound and fall away
> what colored skeins would carelessly
> unwind where I live secretly?[8]

This poem brings to mind the sense of loneliness that comes with being seen not as oneself but merely in terms of one's role in relation to others. It also alludes to the mixed feelings that we have about becoming known by others—it is both what we desire and what we fear. Imagine finding a poem like this written by someone you think you know intimately—your father, mother, spouse, or partner. Would it not cause a shiver to run down your back? And, with a few changes, would it not also describe some aspect of your own life, whether at home or at work? Even though we might find it difficult to put into words exactly what this poem is about, we know that it touches upon something real and very personal.

Many visual artists likewise are astute observers of the human situation. There is a wonderful story about the Norwegian painter Edvard Munch. Even as a child, he gave evidence of having a remarkable talent both as an observer of human behavior and as an artist. One day, while walking through the city, he saw a blind man and, upon returning home, he drew a picture that portrayed something of the essence of what it means to be blind.[9] Pablo Picasso's painting *Guernica* is a world famous rendering of the suffering and the horror of war, just as Auguste Rodin's sculpture *The Thinker* gives form to the tension and concentration involved in meditating on a problem. One of my own favorites is a less well-known piece of art— Sir Hubert von Herkomer's 1891 painting *On Strike*. It shows a family of four standing in the doorway of a modest and dreary dwelling. The father has an expression of determination, wariness, and concern as he looks out at some distant spot. His wife, who is holding their infant, is leaning on him and is looking down at the ground, the very incarnation of discouragement and grief. Further inside the house, we see their adolescent daughter watching them apprehensively. I first saw the painting in 1999, at an exhibition on "Art in the Age of Queen Victoria" at the Frye Museum in Seattle.[10] Of all the paintings there, *On Strike* was the one that left the deepest impression on me. How so? It portrayed so compellingly something of the determination and vulnerability of the working poor as they attempt to stand up for themselves against those who are far more powerful than they. But by no means is this an adequate account of what I experienced when I stood in front of this painting. Words fail me here, although not completely. It might be better to say, quite simply, that this painting made this family of four and the world in which they lived present in the gallery. They were not present in the sense that I felt that they were looking at me; rather, they were portrayed with such care and attention that I immediately grasped something of what life was like for them while almost forgetting myself. At the same time, there was something in this painting that resonated with my values and history. In other words, this painting spoke to me (and, I assume, to others) of something that transcended particular circumstances. Von Herkomer was remarkably effective in communicating his insight into the uncertainty and trials of a family and of creating a visual representation of a specific mood and place such that I was drawn into it. As George Steiner proclaims, in art there is a "shining through" and a creation of a presence that engages us.[11]

However, I do not believe that one has to be an artist to arrive at or express basic insights into human life, or that this happens through some mysterious process that the rest of us do not know about, or that, at the very least, is beyond our capability. But the question remains: how does one arrive at such essential insights, and how does one articulate them in a vivid and fresh way, whether through language or a visual medium? A story told by the poet and writer Kathleen Norris helps to provide an answer to this question. In a chapter on silence, in her book *Amazing Grace,* she writes about the relationship of experience, discovery, and language. While working as an artist in elementary schools in the Dakotas, Norris wanted the children to experience silence. She accomplished this challenging goal by giving them exercises. First, she told the students to make as much noise as possible and to do so to their heart's content. Then they were given instructions for sitting silently, which meant sitting in a relaxed way at their desks. Not surprisingly, it was harder for the students to follow the second set of directions, but after several attempts they reached the point where they were able to sit quietly.

Subsequently, Norris asked the children to describe their experience. She noticed that when the children wrote about making noise, they were entirely unoriginal, relying on stock phrases and clichés. But in writing about silence, a new experience for most of them, they were remarkably creative and articulate. Thus, one little girl wrote, "Silence reminds me to take my soul with me wherever I go," and another said, "Silence is me sleeping waiting to wake up. Silence is a tree spreading its branches to the sun." One fifth grader found silence scary, commenting that "it's like we're waiting for something."[12] Here we see how approaching experiences from a new perspective, seeing them as if for the first time, allows for a freshness of perception and understanding and brings forth imaginative and creative use of language. The children with whom Norris worked were not trained phenomenologists, of course, and yet she helped them to pay attention to their experience and to write powerfully about these experiences. The statements that they came up with—that silence is connected with openness, listening, waiting, and connecting more deeply with the self—are surprisingly perceptive and evocative.

I have given considerable emphasis to how phenomenology intuitively makes sense and how it is connected to everyday experience. This emphasis is warranted because it centers in on what this tradition is all about, and, unfortunately, it is not sufficiently stressed in some of the literature. Furthermore, the foundational texts in existential and phenomenological philosophy are, with some notable exceptions, difficult to read, and this contributes to the tradition generally not being well understood. The works of the German philosophers Martin Heidegger and Edmund Husserl, for example, are dense and in places almost impenetrable without the assistance of a commentary or teacher. Both of these philosophers were academics, and German academics in particular are not known for clear and direct prose. This is neither to disparage the value of their thought nor to minimize the extent of their influence; there is no question that they were profound and creative thinkers. Although their writings are difficult, they have been read by and influenced numerous psychologists and psychiatrists, including Medard Boss, Ludwig Binswanger, and Ronald D. Laing in Europe, and Amedeo Giorgi, Rollo May, and Irvin Yalom in the United States.[13] Phenomenology has also made significant inroads into anthropology, sociology, and other social sciences, and has made important contributions to the development of qualitative research methods in these disciplines as well as in nursing. Yet the tradition continues to be misrepresented in textbooks in these disciplines or is omitted altogether. There are several reasons for this state of affairs; it is not due just to the intellectual difficulties that the philosophical tradition presents. Phenomenology clashes with some of the predominant values of our culture, such as the emphasis on control and efficiency. I will say more about this in the last chapter.

Some of the philosophical writings do provide us with thoughtful reflections on human experience in addition to providing the theoretical justification for such efforts. We see this focus on experience in the writings of the Danish thinker Søren Kierkegaard, who is generally regarded as the founder of existential philosophy. For Kierkegaard it was critical that philosophy address itself to the concrete existence of human beings and that it concern itself not with abstract and purely theoretical questions but basic issues that all of us struggle with, such as making decisions, coming to terms with one's mortality, and finding genuine faith. His concern with understanding existence was greatly at odds with the largely abstract and systematic theory of the German thinker Georg Hegel (1770–1831), the most influential philosopher of his time. As I mentioned in the Introduction, according to Walter Lowrie, one of Kierkegaard's best- known biographers, he was "discontented with Hegelianism because it did not furnish him with *reality* [italics in original]."[14] In other words, reading Hegel's works did little to illuminate or address one's life as a person.

In contrast, Kierkegard's analyses of human relations are relevant to our everyday lives. For example, his insights into despair as an inescapable part of life are psychologically astute. The core of despair, he argues, arises from the insistence that one be something that one is not, or that one not be something that one is. He gives the example of the young woman who is in despair because her beloved was unfaithful to her. At first sight she seems to be in despair over her beloved's betrayal of her. However, Kierkegaard insists, it is more accurate to say that she is in despair over herself. That is, she is left with a self that is without her lover and a self that has been betrayed by him, and it is this self that she repudiates. The authentic way out of this situation is "to will to be that self which one truly is,"[15] or, in other words, to fully embrace who one is. Of course, to understand intellectually the need to embrace or accept oneself does not mean that one can simply do so. As in the case of forgiving another, this is likely to involve a long and arduous process. Along the way one is apt to look for a shortcut, such as falling in love with someone else, wishing that a relationship with a new beloved would enable one to leave behind the "old self."

Roughly sixty years after Kierkegaard's death, the French philosopher Gabriel Marcel (1889–1974) was motivated by his experience working with the Red Cross in France during World War I to develop what he called a concrete philosophy, one that would genuinely take into account the lives of actual human beings. Of his constant contact with the grieving relatives of soldiers who had been killed, he wrote, "Nothing, I think, could have immunized me better against the power of effacement possessed by the abstract terms which fill the reports of journalists and historians of war."[16] In his references to "effacement," Marcel touches upon one of the themes of this book—the overlooking or the denying of the personhood of individuals. We have already seen (in Chapter 1) how Marcel distinguishes between hope and optimism, where the former involves being open to the ambiguity and possibility of life. In contrast, optimists, like pessimists, stand as spectators of life, insisting, on the basis of "conviction," that they know what really is true regardless of what anyone experiences. Marcel's study provides us one of the most profound analyses of hope available anywhere.[17] I first read his chapter on hope while I was working in a psychiatric hospital. It helped me to assess more clearly whether patients who were being discharged were genuinely hopeful about their future or whether their apparent hopefulness was, in fact, very fragile because it was narrowly based on the fulfillment of a particular expectation. For instance, one patient's seemingly positive outlook was based on the belief that he would be able to work for his brother-in-law upon discharge, and that this job would ensure his financial and emotional stability. His exclusive preoccupation with this particular possibility enabled him to keep at bay his terror that he could not manage on his own. Accordingly, we included, in our planning for discharge, discussion of other possibilities for employment as well as identification of community agencies that could help him to deal with his panic if his one "hope" did not pan out.

In the previous chapter, I referred to the psychiatrist George Vaillant, who described his discipline as a poetic science, a concept that has an affinity with Marcel's notion of a concrete philosophy. In other chapters I have mentioned the differences between traditional psychological research methods that are closely based on those of the natural sciences and research methods used by phenomenological psychologists. Given my frequent references to philosophers and artists, and the emphasis on everyday experience as well as on the valuing of persons, the reader may be leaning toward the conclusion that if phenomenological psychology is a poetic science, then it is strong on the poetic and weak on the science side of the equation. It surely is far removed from any conception of science that involves experimentation, objectivity, neutrality, quantification, and control. And if phenomenology is not science, then is it not "unscientific" or merely artistic or literary? It is time to look at the connection between science, psychology, and phenomenology and to examine whether science, as it pertains to psychology, has been defined too narrowly.

The Debate about the Meaning of Science in and out of Psychology

Many students and practitioners of psychology know little about the ongoing controversy about what it means for psychology to be scientific, because most of the literature reads as if this question were settled a long time ago. This is ironic since there has been a healthy debate about the meaning of science with regard to the natural sciences for at least fifty years. The debate is based, in part, on a thorough exploration of the beliefs and practices of scientists and has led to the conclusion that "science" is not an immutable monolith that stands outside of history or culture. This exploration has undermined the assumption that we already know what science is and, therefore, we do not have to think about it. In turn, the whole process of questioning has gradually given rise to some interesting discussions among psychologists about the implications of the new philosophy of science for their own discipline. For instance, in 1983, two psychologists—Peter Manicas and Paul Secord—suggested that psychologists' understanding of science was significantly out of date, a theme that has been echoed in a number of subsequent publications.[18]

The discussion about the nature of science heated up after the publication of Thomas Kuhn's *The Structure of Scientific Revolutions* (1962), a book that convincingly challenged the widely held assumption that science somehow objectively takes account of the world of nature.[19] Kuhn, a well-established historian of science, looked at the actual practice of scientists across the centuries and found that the way in which science was portrayed in textbooks was largely incorrect. His research led him to conclude that science is not characterized by steady, cumulative progress and that scientists do not work from a position of neutrality or objectivity from which they just observe the facts "out there." Instead, Kuhn argued that members of a particular scientific community (e.g., physicists) share a *paradigm*. He defined a paradigm as a set of shared beliefs and assumptions, sanctioned methods of investigation, and guiding examples (what he called *exemplars*) of how scientific problems are resolved. Becoming a member of a scientific community includes learning how to carry out the problem-solving activity of that community and coming to perceive and understand data in a certain way. In other words, one does not become a practicing scientist merely by agreeing to assumptions in science textbooks or following instructions for doing scientific research, anymore than one becomes a competent driver by studying manuals and rules of the road. The practice of science (and this would be true of other professions as well) requires what Michael Polanyi calls tacit knowledge, that is, embodied knowledge that one develops through years of working in the field and through learning from successes and failures.[20] Since much of what practitioners know is tacit, that is, it cannot be fully verbalized, imitating and receiving guidance from the more experienced members of one's community is an essential part of one's socialization as a scientist.[21]

A paradigm, then, is something like a world view. It includes beliefs about the nature of the particular domain that the scientists study (e.g., matter in the case of physics) and how one gathers evidence. In other words, Kuhn challenged the notion that scientists have a neutral stance that is free of philosophical and other assumptions. At certain points in history, some of the observations that a particular community of scientists gather may appear to contradict the very paradigm within which they are working. For example, Kuhn discusses how the Newtonian view of the world of physics was brought into question as it became increasingly difficult to account for new evidence within this framework. Most typically, whenever contradictory evidence arises, scientists will make a concerted attempt to deflect this evidence, initially by dismissing or minimizing its significance, and then, if necessary, by "tweaking" existing theories or explanations to take account of such evidence. Kuhn does not see these responses as problematic but argues that science ordinarily operates from an essentially conservative stance. After all, the paradigm is what enables the scientists to do their work; much effort and thought have gone into its formulation and it would be foolish to question it at every turn. In most cases, these challenges—or anomalies, as Kuhn calls them—can be accounted for within the existing system. Anomalies refer to events that contradict predictions or expectations, and especially those expectations that are foundational. To take an unlikely example, if a person were able to walk through a wall, this would clearly constitute an anomaly for physicists (as well as for the rest of us)! More than "tweaking" of the current paradigm would be required in response.

If anomalies keep cropping up, then the particular scientific community enters into a stage of crisis. The crisis may be resolved through what Kuhn calls a scientific revolution. Someone within the community (e.g., in physics, Albert Einstein and those associated with him) comes up with a new way of looking at the domain with which the community is concerned that appears to account for the contradictory evidence as well as provide a promising direction for future research. Gradually this novel perspective gains support, at least among the younger members of the community, and with time it becomes the ruling view or paradigm.

As one would expect, Kuhn's interpretation of science has given rise to much controversy. Some of his critics fear that his emphasis on psychological, sociological, and historical factors appears to deny that science has universal validity (that is, science is practiced in the same way and with the same results in India as in the United States). And yet it is hard to dispute that science has evolved over time and will continue to do so, and that there are various scientific explanations for a variety of phenomena, both at a given point in time and over time. The distinguished evolutionary biologist Stephen Jay Gould, for one, has rejected the idea that science moves along its own steady route of progress in contrast to other institutions, such as the humanities, that are affected by the "ever-changing winds of social fashion."[22]

A related criticism is that Kuhn's view of science has encouraged the development of "a shallow relativism"[23] insofar as the notion of a paradigm implies that scientists co-create what they study (what one sees depends, in part, on one's perspective toward it) and this interpretation undermines the image of science as testing theories against objective evidence. This criticism distracts us from the broader implications of Kuhn's work, namely, that science does not stand outside of either history or culture. Moreover, in the second edition of his book, Kuhn responds directly to this charge. He states that his position is relativistic only insofar as he rejects the notion that science, over time, gets us closer and closer to some ultimate truth. How would we know that we are getting closer, he counters, since we do not have any independent way of knowing the final truth? In math textbooks, one can always look up the answer in the back of the book, but there is nothing equivalent to this in science. On the other hand, he rejects the suggestion that his view of science excludes the concept of scientific progress merely because it brings into question the idea of final truth. Over extended periods of time, one can see that particular areas of specialization within science become more adept at puzzle solving, and this results in greater accuracy of prediction, increased specialization, and broader range of applications, and thus one can meaningfully speak of scientific progress.[24] My main point here is not that Kuhn is necessarily "right" about every aspect of his thesis, but that his studies have lead to more thoughtful discussions about the nature of science and have paved the way for an examination of the often-hidden assumptions underlying scientific research.

My summary of Kuhn's theory and its critics may appear to be tangential to the question about science and psychology. Yet the opposite is the case since psychologists have borrowed heavily from the natural sciences without engaging in much critical thinking about the implications of so doing. Also, as I have indicated, many psychologists have an outdated and simplistic view of the nature of the natural sciences. Let us start by considering what psychologists mean when they say that their discipline is scientific. (This was briefly discussed in Chapter 3.) Consider the answer provided by Susan Nolen-Hoeksema in her abnormal psychology textbook, a book whose views on science are typical of mainstream psychology.[25] Under the heading of "The Scientific Method," she writes: "Conducting scientific research involves defining a problem, specifying a hypothesis, and operationalizing the dependent and independent variables. Several methods can be used to test the hypothesis." Operationalizing means that you define a concept precisely by indicating how you measure it. The notion of operational definitions comes from a philosophical tradition called positivism, which takes the "hard-nosed" point of view that only that which is directly observable by independent observers exists (at least as far as science and positivistic philosophy is concerned) and that speculation is to be avoided at all cost. In this vein, Nolen-Hoeksema suggests that depression can be defined according to the criteria laid out in the current psychiatric diagnostic manual (the DSM-IV) or according to a score on a test that purports to measure depression. With these definitions, one could argue, at least there is clarity about what one means by depression.

Defining one's terms precisely is only the beginning point, however. The next step is to formulate a hypothesis and then test it through the use of specific methods. This testing of carefully defined hypotheses through approved methods is what she (and many others) believes legitimizes a discipline's claim to be scientific. The three methods that she lists are case studies, correlational studies, and experimental studies. She comments on the advantages and disadvantages of each method. The case study method deals in depth with the individual, but it is hard to generalize from one case to another because each person is unique; cases also lack objectivity because you rely on the perspective of the person telling the story (the patient) and that of the listener (the therapist). On the other hand, laboratory studies of human behavior enable researchers to control what happens and allow them to draw conclusions about cause-and-effect relationships, at least with respect to this controlled situation. One could, for instance, assess whether short-term exposure to full-spectrum light reduces depression in people with Seasonal Affective Disorder (SAD). The problem is that laboratory experiments take place under artificial and limited conditions, with obvious ethical and practical constraints on what researchers can do. For ethical reasons, researchers are not going to instruct the parents of fifty identical twins to favor one over the other in order to determine what effect this would have over a ten-year period on the twins' emotional adjustment and related factors. The third method is correlational in nature and focuses on how two or more variables are related to one another in terms of patterns. If one variable decreases (amount of sunshine), does the other also increase (depression in people diagnosed with SAD)? This method allows one to collect data in the world at large (as opposed to the laboratory), but it is hard to know what the specific relationship is among the variables that are correlated. "Correlation does not equal causation" is the mantra of every textbook in the social sciences and in statistics.

As is true of many psychologists, Nolen-Hoeksema takes the position that first-hand accounts are not reliable, that the best psychological data is that collected by impartial observers, and that one should account for human behavior in terms of explanations that make sense of behavior without considering the acting person's point of view. This is the same view of science and scientific method to which I was exposed in my introductory psychology class in the mid-1960s, and it is a view that a number of psychologists have come to believe is untenable. Manicas and Secord, for example, have argued that one cannot explain human behavior without reference to "ordinary descriptions of behavior and experience."[26] Admittedly, today's version of scientific psychology is more sophisticated than was the case forty years ago. Also, there are new methods that are used by biomedical researchers (computerized tomography, positron-emission tomography, and magnetic resonance imaging) to which psychologists give a lot of credence. What is often forgotten, however, is that these are *not* innovative *psychological* methods—they do not focus on human behavior and experience.

Paradoxically (and this is noteworthy), Nolen-Hoeksema also emphasizes the intersection of science and humanity, giving the reader accounts of the "subjective" experience of the mentally ill. Of course, this is indispensable for any textbook that has an applied focus, but her book does this more extensively than most texts. Nonetheless, it is still a strange state of affairs that the science and the humanity dimensions are treated as two separate tracks. Furthermore, the case-study approach, which is evaluated in relatively critical terms from a science perspective, is then brought in as indispensable in order "to highlight the personal experiences of people with mental disorders, to give students an appreciation of their suffering and courage, and to help students understand their personal encounters with psychopathology."[27] So, while the symptoms of disturbance and the brain structure and functioning of those with diagnoses are studied scientifically, the courage and suffering of those with these disorders are on the other side of the fence because these are "subjective phenomena" that cannot be studied reliably!

Without someone's personal encounter with "psychopathology," there would be no research, because all research presupposes observations of people who are disturbed as well as these people's descriptions of their own experience. Or, to use Fischer's terminology, one cannot have secondary data unless one first has primary data. Fischer, in turn, is indebted for this distinction to the work of Husserl, whom I mentioned earlier. Toward the end of his life, Husserl developed the idea of the *lebenswelt,* or life world. This is the realm of immediate human experience existing prior to the abstractly conceived world of the natural and the social sciences, including psychology.[28] Herbert Spiegelberg, a leading phenomenological philosopher, has defined the life world as "the encompassing world of our immediate experience."[29] The life world is the realm scientists depend upon even while they also tend to negate or overlook it. To be more specific, one cannot begin to discuss criteria for the diagnosis of depression or construct scales to measure depression until one has looked carefully at the behaviors and experiences associated with what we call depression. The experiences and behavior precede attempts to measure or explain them, and yet, somehow, explaining or measuring is thought to be more scientific than studying the phenomena directly? Surely something is amiss here.

Psychology as a Human Science

The history of psychology is more ambiguous than I have suggested so far. Although the majority of psychologists have accepted the natural sciences as a model for their own discipline, there have also been numerous thinkers along the way who have been critical of psychology's emulation of the methods of physics and chemistry and have articulated alternative visions. Even the so-called father of modern psychology, Wilhelm Wundt, who started the first psychological laboratory in 1879, had a view of psychology that would disturb many a modern psychologist. He was strongly opposed to the separation of psychology and philosophy and thought that experimental methods were only adequate for studying simple perceptual processes. Higher mental processes should be studied using more qualitative methods, and philosophical reflection was by no means to be avoided in dealing with psychological issues, according to Wundt.[30]

The eminent American psychologist and philosopher William James (1842–1910) had a complex vision of psychology. He was described as a radical empiricist because of the value he placed on direct experience, and he was also quite critical of the materialism and reductionism of late nineteenth century scientific philosophy. Moreover, his famous study, *Varieties of Religious Experiences,* provided an in-depth analysis of religious phenomena that relied on first-hand accounts. James, who had been trained as a physician and had taught physiology and anatomy, suggested that psychology should be considered a natural science. But his overall perspective on psychology was that it had to do justice to the complexity and mystery of what it means to be a person.[31]

The call for psychology to do justice to the complexity of human existence was taken up in a vigorous and systematic manner by Giorgi, in his groundbreaking 1970 book *Psychology as a Human Science: A Phenomenologically Based Approach.*[32] Giorgi had received his doctorate in experimental psychology and had gone on to work as a project director in the area of human engineering. Both as a student and as a practicing psychologist, he had been dissatisfied with psychology's ability to address human problems. When he returned to the academic world, he began to develop a model for psychology different from the one he had been exposed to during his graduate education.[33] His book provides a powerful vision for those psychologists who want to study human beings in human terms and who have recognized that psychology's adherence to a natural science model gets in the way of this fundamental goal. But Giorgi does not suggest that psychology conceived as a natural science is without value. Instead, his point is that this model does not allow for a study of human beings that is faithful to the experience and behavior of actual human beings. For one thing, mainstream psychology gives priority to the quantitative over the descriptive. Overall, Giorgi critiques psychology for not doing justice to complex and subtle human phenomena. Keep in mind that at the time that Giorgi wrote his book, psychology ignored topics such as forgiveness. More recently, this phenomenon is being given attention by psychologists but, as we have seen, without doing justice to its paradoxical and subtle nature because of the preoccupation with quantification and the experimental method. Moreover, Giorgi has pointed out that psychology has not developed holistic methods that would allow it to study human experience on its own terms. The primary problem is that psychology has uncritically adopted the philosophy on which the natural sciences are based, or at least the one on which the discipline was based a hundred years ago when it had its origin. Following from this argument, he states his fundamental position: one should not have to choose between a psychology that does justice to human experience and a psychology that is scientific. In his review of the history of psychology, he shows that were a number of thinkers (such as James) who called for a psychology that was descriptive or interpretive rather than just experimental or explanatory in its orientation.

One of the key concepts in his book is that of *approach,* which is similar to, but somewhat broader in scope than, Kuhn's concept of paradigm. Throughout *Psychology as a Human Science,* Giorgi draws on Kuhn's research insofar as he opened up the whole question of how science actually works when one looks at its practice in a historical perspective. Giorgi defines approach as "the fundamental viewpoint toward man [*sic*] and the world that the scientist brings or adopts, with respect to his work as a scientist, whether this viewpoint is made explicit or implicit."[34] On the basis of his study of phenomenological philosophy, and especially the work of Husserl and Merleau-Ponty, Giorgi tackles foundational issues in psychology, issues that psychologists have avoided since the discipline's alleged break with philosophy. This is a large part of where the problems in psychology come from, Giorgi argues. Because psychologists have been unwilling to look at their foundation, that is, their basic world view or paradigm, far too many questions about appropriate methods and about the definition of psychology's subject matter have been answered without adequate reflection. To assume that the experimental method is the best way to study human phenomena, for example, is to uncritically subscribe to a positivistic philosophical view of human beings and of human knowing. What the natural scientific philosophical view of human forgets, asserts Giorgi (following Husserl), is that the life world is primordial, whereas the world that the natural sciences study is constructed on the basis of our experience of the life world. To put it more succinctly, the activities of scientists presuppose the everyday world. This is the basic point I made earlier with regard to research on disturbed behavior. It is the experience of the life world that psychology should investigate, using whatever methods are appropriate. In setting out his positive vision of psychology as a human science, Giorgi suggests that it should be based on three suppositions: fidelity to the phenomenon of humans as persons (rather than as organisms, for example), special concern for uniquely human phenomena (disillusionment and forgiveness, along with love and creativity, would fall into this category), and recognition of the primacy of human life as relational.

In terms of its origin, the word "science" comes from the Latin word *scientia,* "to know," and thus the word "scientist" refers to someone who has knowledge, especially of a technical kind.[35] So, at first sight, Giorgi's attempt to redefine psychology as a human science does not present any problems because, with respect to its etymology, "science" has a broad meaning. Yet redefining psychology is an uphill battle. Advocates of funding for research in psychology from government sources (e.g., the National Institute of Mental Health) argue vigorously that psychology is as scientific (and in roughly the same sense) as related disciplines such as medicine and biology.

In the popular imagination, "science" connotes people in white lab coats conducting experiments and coming up with conclusions that are indisputable. The concept of "scientific evidence" has, as we saw in Chapter 3, connotations of status and certainty. Some of the social sciences, such as political science and economics, have frequently been the object of scorn, both of the public and of those within the natural sciences, because their practitioners seem to be unable to agree among themselves what is going to happen in the next election or in the stock market the next year. For every economist one can round up who argues that a specific budget policy will create jobs, one can easily find another who will argue that the opposite will happen. Sylvia Nasar, Nobel Laureate John Nash's biographer, has described the opposition of the natural scientists in the Royal Swedish Academy of Sciences to the idea of a Nobel Prize in Economics. Their objection was that in economics "one could not point to scientific progress, a body of theories and empirical facts about which there was certainty and near-universal agreement."[36] Psychology and psychiatry have been subject to similar critiques and are in some ways more vulnerable because they are often called upon to predict the behavior of individuals rather than the behavior of large groups of people such as voters or investors. Psychologists and psychiatrists have modest success, at best, in predicting, for example, whether a person will do harm to the others or attempt suicide.[37] Of course, one of the reasons that this sort of prediction is so difficult is that people's behavior depends upon their circumstances, and so, while one can carry out a thorough assessment of individuals, one cannot know what situations they will be faced with in the future.

When one looks at a current dictionary to find out how "science" is understood in contemporary society, one finds a relatively broad array of meanings. *The Random House Webster's College Dictionary* provides the following definitions:

1. A branch of knowledge or study dealing with a body of facts or truths systematically arranged and showing the operation of general laws.
2. Systematic knowledge of the physical or material world gained through observation or experimentation.
3. Any of the branches of natural or physical science.
4. Systematic knowledge in general.
5. Knowledge, as of facts or principles; knowledge gained by systematic study.
6. A particular branch of knowledge.
7. Any skill or technique that reflects a precise appreciation of facts or principles.[38]

Of these definitions, 2 and 3 refer specifically to the natural sciences. The first interpretation is ambiguous since it is not clear whether any of the social sciences can justly claim that they have identified "general laws," or that this concept can appropriately be applied to the behavior of human beings.[39] Definitions 4 through 7 would include the idea of psychology as a human science with an approach and methods of its own rather than ones based on the natural sciences. It is not my intention here to settle the controversy regarding psychology's status as a natural or a human science, but to outline some of the basic issues. Above all, I want to show that there is a coherent rationale for asserting that phenomenological psychology is a scientific approach. Those favoring the natural scientific model for psychology insist that psychology is scientific because it uses the established methods of science.

I argue, as do Giorgi and Elio Frattaroli (the psychiatrist I referred to in Chapter 4), that the methods one uses be appropriate to the subject matter, and if the subject matter is human experience, then laboratory and correlational methods are inadequate. In fact, these methods are not sufficiently empirical. That is, the data collected with these methods do not include descriptions of experience, but measurements of "variables." The researcher is thus distant from what he or she tries to account for, and the explanations that are offered are speculative. They are speculative because the researcher is guessing as to what "causes" the participants of the study to act as they do, without directly asking what was going on for them during the experiment or the study. I am not suggesting that experimental research participants (or any of us, for that matter) can necessarily fully account for why they do what they do. However, without information about their perception of the situation they are in, any explanation that a researcher offered is guesswork at best. As the social psychologist Kurt Lewin asserted years ago, accounting for the behavior of a student objectively requires that one take into account the classroom as it exists for him or her, not as it appears to the adult who is studying the student.

Phenomenological Methods

What do phenomenological psychologists do when they do research? In the previous chapters I have briefly addressed this question, indicating that one starts with interviews or written accounts and then carefully reflects on them, attempting to arrive at a description that articulates the basic structure or essence of the experience being studied, be it disillusionment, anger, or falling in love. At this point I will discuss phenomenological methods in greater detail, although it is not my intention to write a guide for researchers. There are plenty of such guides available.[40] To show that there is a range of methods in use, I will discuss two that are significantly different and then turn to what they have in common. This approach is itself characteristic of phenomenology. That is, as in the first three chapters, I presented various descriptions of a phenomenon, such as disillusionment, reflecting on them to discern which of their features are true of any experience of being disillusioned and which features are specific just to some people's experience. Thus, an essential feature of disillusionment is that we are not only profoundly disappointed in that person but the very meaning and direction of our life and our relationship to the other simultaneously come into question.

Often the person with whom one is disillusioned is older (e.g., a parent or mentor), but this is not an essential theme or constituent of this phenomenon. For example, several people were disillusioned by a friend or lover. This effort to identify what is essential to a given phenomenon is technically called the *eidetic reduction* (from "eidos," meaning "essence"), a concept developed by Husserl.[41]

As I indicated earlier in this chapter, artists may grasp the essence of a phenomenon intuitively, but within the phenomenological tradition there are basically two systematic ways of reaching this goal: free imaginative variation and empirical variation. The former, as the name suggests, relies on the imagination. One takes a particular phenomenon or object and keeps varying it imaginatively to arrive at what is essential to it. The example that is often used in clarifying this process is a chair. What is the essential nature of a chair? Does it have to have four legs? Obviously not. Does it have to have more than two? No. And so on. Eventually one arrives at the essence of the chair as that piece of furniture designed for us to sit on—it enables us to sit. With empirical variation one looks at a number of actual descriptions of a phenomenon and one discerns what they have in common that is also essential to the phenomenon. This process is not as esoteric as it sounds. If a group of people share stories of a particular experience (e.g., losing a parent), that which is in common will typically "jump out" at them. That is, the members of the group are not necessarily making a conscious effort to seek out the "essence" of what is at the core of these stories, and yet, as one listens to story after story, the commonalities start to register. None of this is to suggest that these essences or general patterns are independent of their cultural context or that we can describe them in a final or definitive way.

This first method I have selected is the best known and most widely used in phenomenological psychology and was developed by Giorgi. It provides a sequence of steps for the individual researcher. The second method, less well known, was developed at Seattle University by my colleague Jan Rowe and me, as we worked with a small group of graduate students; we used this *dialogal phenomenological* method in our research projects on forgiving another and forgiving oneself, as well as in our current research on the experience of hopelessness.[42] Our method does not involve following a preconceived structure as does Giorgi's method, but places its emphasis on working collaboratively to decide what steps need to be taken given the particular phenomenon that is being studied.[43] It is obvious why I have selected the first method. The reason for my selection of the second method is, as I have implied, that it at first appears to be almost entirely different from Giorgi's method, and thus the exploration of what these methods have in common becomes more interesting and informative. Also, in the interest of full disclosure, I have to confess to being partial to the dialogal method as a valuable and creative way of carrying out phenomenological research. My own experience is that working with others whom one has come to trust makes it possible for one to reach a level of creativity and understanding that exceeds what the individual is able to accomplish on his or her own.

With any phenomenological study, one first has to decide what experience or phenomenon one wants to study and, second, what question or questions one is going to ask people to respond to, either in writing or in an interview. This is not necessarily as straightforward a step as it might seem. In his early study of learning, Giorgi used the following question: "Could you describe in as much detail as possible a situation in which learning occurred for you?"[44] Note that the question asks the person to describe a particular experience, but that the choice of what is defined as learning is left to the individual. That is, the question is both focused (on a particular type of experience) and open-ended in the sense that the person is giving full rein to tell his or her story. The woman who was interviewed described learning to see the vertical and horizontal lines in a room with the aid of a friend who was an interior decorator. With this new awareness she was able to rearrange her living room so that it looked better.

Once you have the description—in this case the transcription of the interview—then what do you do? Giorgi's method involves the following seven steps, the first of which is obtaining the description. Some phenomenological researchers, such as Fischer and Frederick Wertz, who have drawn upon Giorgi's method in studying the experience of being criminally victimized, suggest that the researchers also write their own description at the outset. Doing so allows for reflection on one's own preconceptions and gets one more closely in touch with the phenomenon. Thus, by writing about being criminally victimized and remembering how difficult such an experience was, the researcher is better prepared for the delicate task of interviewing others who have been victimized.[45]

The second step is to read (and reread) descriptions collected from research participants in an attitude of openness, not trying as yet to get an answer to the research question (e.g., what is the nature of learning?). Instead, the goal is to attempt to understand what is going on from the person's point of view. This kind of reading requires that one attempts, imaginatively, to place oneself in the situation of the other, rather than, for example, comparing the other's reaction to what oneself might have done. The more dispassionate aspect of this process comes with noting what the person says, rather than disbelieving or believing it to be "objectively" true (e.g., the person might say, "If it hadn't been for my wife's strict upbringing, this would never have happened"). Then, having read the description as a whole, the third step involves breaking it into "meaning units," a concept analogous to a complete thought. Fischer and Wertz explain: "The criterion for a unit was that its phases require each other to stand as a distinguishable moment in the overall experience."[46] The following is an example from Giorgi's study:

> I found out what was wrong with our living room design: many, too many, horizontal lines and not enough verticals. So I started trying to move things around and change the way it looked. I did this by moving several pieces of furniture and taking out several knick-knacks, de-emphasizing certain lines, and . . . it really looked different to me.[47]

Deciding where one meaning unit ends and another begins involves judgment, obviously, and it is not as if there is one right way to arrive at this. Meaning units may be relatively short (e.g., several sentences), or the length of an extended paragraph. In my experience, however, novice researchers tend to create too many meaning units. Then, in the fourth step, one states the primary theme of each meaning unit; the theme may be implicitly or explicitly present in the original. For the above meaning unit, Giorgi wrote, "S[ubject] found too many horizontal lines in living room and succeeded in changing its appearance."[48] During the fifth step the researcher looks at the themes that have been drawn from the description and attempts to see how these themes address the research question or questions. I will clarify what this means with reference to Giorgi's study. In this study, Giorgi posed two questions, but for the sake of simplicity I will just focus on one: what is the nature (or structure) of learning? The following is his transformation of meaning unit 3 to respond to this question:

> The subject recognizes what is "Wrong" with her room and rearranges it according to her perception of the lines. Once again a real change is evident because the furniture is rearranged and room appears different to S.[49]

As you can tell, the formulation of each meaning unit is in the context of its place within the overall story or description. In the above transformation Giorgi implicitly refers back to what this participants had been told by her friend about vertical and horizontal lines.

The sixth step involves writing a *situated structure*. This is a summary of what is essential to the particular description, couched in terms that retain the specificity of the experience. Anything that is not essential to the experience as it occurred for this person is left out, and yet the essential dimensions are still presented in their concreteness. In this case, the situated structure referred to how this woman gained an awareness of the importance of horizontal and vertical lines from a friend, started seeing her own living room in these terms, and used this new awareness to rearrange the room so that it looked better to her. It also looked better to her husband, but he could not account for the difference.

The final and the most difficult step involves coming up with the *general structure*. Here one asks: what is at the core of this experience as an experience of learning? Arriving at the general structure is certainly easier if one has descriptions from several research participants with which to work, because one can then ask, what do all of these accounts essentially have in common? On the basis of his reflection on this one description, Giorgi came up with the following:

> Learning is the ability to be present to, or exhibit, the "NEW" according to the specific context and level of functioning of the individual. This awareness of the "NEW" takes place in an interpersonal context and it makes possible the appreciation of a situation in a fuller way, or the emergence of behavior that reached a different level of refinement in a sustained way or both.[50]

In this case the appreciation of the situation refers to the awareness of the effect of horizontal and vertical lines, and the behavioral aspect refers to the person's ability to change the room to her liking on the basis of the new awareness. In the case of learning to drive, for example, the student driver might gain an awareness of how one has to work the clutch and the gas pedal in concert, and the behavioral aspect would involve the implementation of this understanding, that is, letting the clutch out gradually while pressing lightly on the accelerator so that the car starts smoothly.

I will return to Giorgi's method after I have discussed the dialogal phenomenological method. However, it is important to recognize that this method, while it involves specific steps, cannot be implemented mechanically. In other words, it is not a technique in the sense meant by William Barrett. In the previous chapter I mentioned that Barrett defined a technique as a standard method that can be taught, and it can be taught because its steps can be precisely specified. Using Giorgi's method involves judgment and imagination, and there is a sense in which one does not really appreciate the method until one has worked with it for a while, ideally with the guidance of an experienced phenomenological researcher. As Kuhn pointed out, you do not become a competent member of a scientific community just by reading its texts and manuals.

The dialogal method entails having a small group of researchers—typically four to six—working together. Whereas in Giorgi's method the researcher follows a predefined set of steps, in this method decisions as to how one should proceed are made by the group. Perhaps the best way to introduce this method is by looking at the nature of dialogue as it happens in everyday life. A dialogue is simply a focused conversation, whether with one person or with several people, that leads to a deeper personal understanding of, or insight into, an important aspect of our lives. For example, you might have had an especially meaningful discussion with someone about the nature of friendship, leading you to the realization that while you might take your friends for granted, friendship is an especially important and sustaining aspect of your life. These types of conversations take on a life of their own and lead us to places that we had not anticipated; in other words, they involve discovery. The German philosopher and Heidegger's former student Hans-Georg Gadamer has noted that "we say that we 'conduct' a conversation, but the more fundamental the conversation is, the less its conduct lies within the will of either partner."[51] That is, once we get involved in the discussion of a topic, the conversation takes on a life of its own, and it is as if the topic itself, rather than just the individuals who converse, guides the discussion.

In late 1984 Rowe and I, along with four graduate students, began the study of forgiving another that was mentioned in Chapter 3. We did not plan to "invent" a new phenomenological method, but as we reflected back on the process that had evolved during this project, it became clear to us that this is in fact what had happened.[52] Perhaps starting out without a clear idea as to how we would proceed, other than having a commitment to working collaboratively, enabled us to find a new way to do research. In addition, the topic of forgiveness helped to give a certain collaborative ethos to the group because of its inherent emphasis on overcoming injuries and moving toward reconciliation.

We call this method "dialogal" because, at its core, it is characterized by two simultaneous levels of dialogue: among the researchers and between the researchers and the phenomenon being studied. Faithfulness to the phenomenon is fostered through open conversation among the researchers in relationship to the data—in the form of descriptions or interviews—and through careful consideration of the various perspectives of the members of the research group. Dialogue is the basis for every step of the project: making decisions about process, dividing up tasks, reviewing pertinent literature, and interpreting data.

The success of this method depends on the relationship of the researchers to the phenomenon being studied and the relationship among the researchers. It is essential that the researchers be fully committed to studying the phenomenon under investigation, because the ongoing focus on the phenomenon and the concern with understanding it as fully as possible are what give structure and cohesion to the group process. This does not mean that there are not secondary agendas of a personal nature at work, such as a desire for meaningful contact with others or scholarly achievement, but the primary focus should be on arriving at a shared understanding of the phenomenon. Thus, in interviewing graduate students who want to join one of our research projects, Rowe and I try to determine whether the students are genuinely interested in learning about the phenomenon, as opposed to being strongly invested in a particular theory about the topic or viewing the research project as an opportunity for resolving personal issues. The phenomenon is brought into focus through the researchers' sharing descriptions of their own experience of the phenomenon—be it forgiving another or experiencing hopelessness—and through the interviews that are conducted later in the process to get descriptions of the phenomenon from "subjects" outside of the group. That is, the data for the study include accounts written by the researcher (at least in those cases where the researchers have first-hand experience of the topic at hand) and accounts provided by research participants.

This ongoing focus on the phenomenon allows it to take on a life and a presence within the group rather than being something abstract and "out there." Indeed, there are times when the topic is palpable within the group. As the psychologist Eugene Gendlin has written, "If experience appears, it talks back."[53] This is not something mysterious, but something with which we are already familiar in everyday life. Imagine, for instance, a family reunion from which a favorite uncle is absent. As those who have gathered together talk about him, it is almost as this uncle becomes present, and as one person tells a story about him, this brings to mind another story for someone else. The presence of the phenomenon in the research group is even stronger because the group has a number of descriptions in front of it and constantly refers back to these descriptions. It is this kind of intense focus that gives structure or direction to the research. The topic under consideration directly affects how people work together, and this creates its own challenges. In our research study of despair as it is experienced in everyday life, we found that it was harder to stay focused on the topic than had been the case when we studied the experience of forgiving another. The study of despair led us to spend more time giving support to each other, and there were times during the early phase of this project when we seriously questioned the feasibility of our study. It took us some time to realize that these responses and these doubts were signs of our involvement with the topic rather than signs of failure, just as shortness of breath may be an indicator that one is getting closer to the peak of the mountain. At some point, one of our members noted that our group seemed as confused about the direction our research ought to take as the people we interviewed were about making sense of their lives in the middle of their experience of despair. This comment helped us to understand how our process reflected the nature of the topic we were studying.

To work in this way there must be a growing trust among members of the group, a trust that each one's experience will be treated carefully and not judged, a trust that members will be open to one another's words even when their relevance is not evident, and a trust that allows the researchers to feel valued and to be able to count on one another. At the start of the process, it is enough that the group members share the hope that this is possible, because such trust, faith, and commitment require sitting with one another for some time before these can develop. Michael Leifer, who was a member of the first forgiveness group, transcribed the meetings of our group and another research group. The second group was led by one of our colleagues, George Kunz, and focused on the topic of humility.[54] Leifer's study helped us to understand the gradual way in which the group process develop over time. Drawing upon Gadamer's philosophy, he identified three levels of dialogue: preliminary, transitional, and fundamental. The basic idea is that initially the researchers talk somewhat abstractly about the topic being studied as they gradually get a better sense of who the other members of the group are. Then, as trust grows, the discussion becomes more experiential, and there are more references to actual experience of the phenomenon. The fundamental level of dialogue refers to those times when there is a primary focus on the phenomenon being studied and continuity of conversation as the researchers listen carefully to, and build on, what the others are saying. These three levels are by no means mutually exclusive, but as the researchers continue to meet, more time is spent in fundamental dialogue.[55]

Insofar as focus (or structure) and trust are present, these dialogal research groups show creativity both in how they approach the topics under study and in how they articulate their understanding of the topic; the way these collaborative groups function certainly provide support for the proverb that two (or six) heads are better than one. First of all, one does not have to rely on oneself alone to recognize one's own prejudgments. Instead, through a multiplicity of perspectives arises the possibility of seeing the phenomenon in new ways. Similarly, it is easier for a collaborative group than for an individual researcher to move past obstacles or impasses, because what one person cannot see or imagine, another may. Also, with a group, it is easier to find words to describe what one has come to understand about a phenomenon. To begin with, there is, as I have mentioned, the sense of the presence of the phenomenon among the researchers. While it is true that being so close to a phenomenon can hinder seeing or reflecting on it, with a variety of perspectives and ways of saying things, sooner or later we found ways to say something. One word leads to another, just as the first drop of water is the beginning of a stream. Once, when even a few words have been spoken, they can be tested out for fit, and one can come up with more. This process is an intermingling of receptivity and creativity, of discovering truth and creating truth. As the philosopher George Gusdorf writes, "Speaking is not merely a means of expression, but a constitutive element of reality."[56]

All of this presupposes, of course, that the group members are able to use disagreements and diversity of viewpoints constructively, that is, to treat these as a basis for further exploration and not take them so personally. This capacity for dealing with differences is certainly one of the clearest indicators of the degree of trust established in the group and of its potential effectiveness. If the researchers take one another and their project seriously, they will hear one another out and patiently explore the various hunches and insights that come up in their discussions. The key principle in the dialogal approach is not that the group members find a workable compromise, but that they reach the point where all of them can affirm that an analysis or interpretation does justice to the data.

The differences between these two methods are numerous and, for the most part, obvious. Giorgi's method involves a single researcher who goes through a number of prescribed steps. The dialogal method takes a small group of researchers who have to deal with a great deal of uncertainty because they have only general principles to fall back on (stay focused on the phenomenon, listen carefully to what others have to say, and learn to trust your own experience and the group process) and make up the next step as they go along. A reader or reviewer of a study that follows Giorgi's method can examine in detail the steps in the process, whereas, with the dialogal process, all that can be directly checked are the transcripts of the data, the researchers' own accounts, and then, finally, the analyses of the data. In some cases, there may be an intermediate step—a summary of the first interview that is presented to the interviewees as part of a second, follow-up interview. Overall, it may not be so clear how the group moved from data to analyses, although the analyses typically includes numerous quotes from the transcripts so that the reader can compare the findings and conclusions with the specific descriptions of the phenomenon.

In spite of these (and other) differences, there is also much that these methods have in common. Above all, both methods are structured to foster faithfulness to the phenomenon being studied. The Giorgi method does so by setting up steps each of which requires an ongoing return, at various levels, to the descriptions that have been collected. In the dialogal groups the focus on descriptions is what gives structure and direction to the research project. To move forward, a group has to clearly identify the phenomenon of interest and remain focused on it. The very notion of the presence of the phenomenon as a partner in dialogue, a notion that is based on the actual experience of researchers using the dialogal method, emphasizes how the means of studying a particular experience is responsive to the very nature of that experience. Both of these methods are empirical in that they are based on descriptions (or observations) of experience.

Of course, one can focus on an experience even while one imposes upon it (often without knowing it) one's preconception of what it is and how it should be understood. As a number of philosophers (e.g., Heidegger and Gadamer) have pointed out, one inescapably proceeds from some already existing perception or *preunderstanding* of a particular question or issue. If one did not have some notions about the issue, one would not attend to or ask questions about it. The point is not that one's preunderstanding is necessarily wrong (although obviously it is incomplete), but that it is important to become cognizant, as far as possible, of what it involves, so that one can be mindful of its influence and limitations. Stephen Jay Gould is emphatic that scientists necessarily bring a bias to their studies. "The peculiar notion that science utilizes pure and unbiased observation as the only and ultimate method for discovering nature's truth operates as the foundational (and, I would argue, rather pernicious) myth of my profession."[57]

He goes on to argue (in line with Kuhn's thesis) that scientists look at data from a theoretical perspective and have preferences in terms of what they hope to find, and that, being mindful of this reality, most scientists subject these assumptions and expectations to a rigorous test. Phenomenologists, almost by definition, are also highly mindful of preconceptions and biases in their approach to the study of experience. If one is to approach phenomena in an attitude of wonder, attempting to see them as if for the first time, one has to find some way to set aside one's usual expectations and assumptions. The technical word for so doing is *bracketing*. It is not that one necessarily judges one's assumptions to be false (although this will be true of some assumptions), but that one temporarily takes them out of play. Bracketing is a process, not something that one does all at once. To begin with, all of us have numerous assumptions about any number of things, and most of them we do not recognize as assumptions. The discovery of one assumption often leads to the discovery of another.[58]

The most fundamental assumption that Husserl identified was that of the *natural attitude*. By this he meant our assumption that what we experience is simply a reflection of what is there. To put it simplistically, if we experience a man as unpleasant or a landscape as appealing, we take it for granted that this is the way the man is and the way the landscape is. What you see is what is there. The natural attitude, the attitude with which we live on a daily basis, overlooks or forgets that as human subjects we co-create or co-constitute what we experience, that our approach or perspective on things, individually or culturally, helps to determine the nature of our experience. However, this is not to be confused with the "New Age" notion that we create our own reality. Since it is critical to make the distinction between co-constituting and creating one's own reality, let me briefly explore a few examples. Personally, I find the wide open spaces such as the ones that are characteristic of much of Montana oppressive. Many other people experience a sense of liberation when they are in those kinds of open spaces. But neither those who enjoy or are troubled by vast expanses "create" their vastness. With interpersonal relations, the matter becomes more complicated, so I will just allude to some of the issues involved. As we saw in Chapter 1, one's experience of another can change significantly as circumstances and one's own attitudes change. Thus, a man I initially experience as unpleasant, perhaps because he is often sarcastic and curt, may subsequently be someone I recognize as fundamentally compassionate when he empathizes with my loss of a friend. In turn, it may well be that the directness with which I speak of my loss is what brings out his compassion. But none of this is to suggest that the people we relate to are creations of our own mind or that any or all interpretations of what they are like are as true or false as any other interpretations. Some of our understandings of others turn out to be woefully inadequate.

So let us look at the notion of bracketing more concretely by returning to phenomenological research. First of all, the phenomenological researcher suspends the natural attitude by looking at experiences and setting aside the concern with whether and to what extent they reflect some reality beyond the experience itself. For example, William Fischer, a psychologist at Duquesne University, has studied the experience of being anxious over a number of years, using a revised version of Giorgi's method. Thus, in his studies, Fischer focuses on what people's experiences of being anxious actually are, and he gets at this by asking them to write descriptions of a time when they experienced being anxious.[59] Similarly, as you may remember, Giorgi asked his subjects to give descriptions of a time when learning took place for them. In approaching this topic, William Fischer puts aside or brackets some of the assumptions that are made in psychology about "anxiety." One of these assumptions is that anxiety is a "thing" or a force (note that he asks people about their experience of being anxious), that it is an unwarranted fear (this is the judgment of an observer who does not consider what a particular situation means to the person who is anxiously responding to it), and that it is somehow divided into situational, emotional, and bodily components. Moreover, he also places aside everyday assumptions that being anxious is "bad," or that it is either something that people could overcome if they only wanted to, or, on the contrary, that it is something that comes upon people like a thunderstorm. Although I will not give a detailed account of his findings, I will at least, give a brief indication of the direction of his analysis:

> An anxious situation arises when the self-understanding in which one is genuinely invested is rendered problematically uncertain, and hence, possibly untrue. Two variations of this situation may be delineated: In the first, an essential constituent of that self-understanding, one that expresses one's iden tification with a state of affairs that one is endeavoring to realize, for example becoming a PhD candidate in one's graduate program, is now experienced as possibly unattainable, and thus the entire self-understanding that one is living is called into question; in the second, a meaning that one is living as either never-to-be-true of one or no-longer-to-be true-of-one, for example, being someone who "gives in" to the desire to masturbate, has emerged as possibly (still) true, thereby undermining the self-understanding, that is, at least in part, founded upon its absolute exclusion.[60]

This analysis refers back to the common themes that emerged in the descriptions that William Fischer gathered. His analysis (as is true of phenomenological analyses in general) goes beyond the actual words that were used in the descriptions as it moves toward identifying and articulating the implicit dimensions of the experiences that his respondents wrote about. Obviously, a study such as this is not "presuppositionless," as if this were possible. Phenomenologists assume that people are able to describe their experience with some degree of reliability or completeness and that there are commonalities across various people's experiences of phenomena, such as being anxious. In addition, William Fischer's analysis carries within it, at least to a limited extent, the influence of psychological and philosophical theory. This is evidenced by his use of words such as "self," "self-understanding," and "identification." No doubt one could write an analysis using different words, but one cannot escape traditions altogether (a point made by Gadamer), although one can become increasingly cognizant of, and thoughtful about, their influence. And it is an open question to what extent this analysis of being anxious would apply to cultures markedly different from that of the United States. Psychological research, be it phenomenological or traditional in nature, does not yield universal or eternal truth. However, two of the most pertinent evaluative questions one could ask in this context are whether William Fischer's research (following Karp's criteria) is evocative of our own experience of being anxious and whether it leads us to understand it in a new way. These are questions that I would encourage you to ask on the basis of your own experience of being anxious.

With this discussion of bracketing and biases in place, let me return to Giorgi and the dialogal phenomenological methods and consider how they deal with these issues. As we have seen, Giorgi's method requires a series of steps, starting with reading the text, be it a written description or a transcript, in an attitude of openness. Initially, one stays very close to the words of the text and to the purpose of the changes and transformations that are undertaken after one has broken the text into meaning units. The changes are toward greater generality and focus on the basic research question, but the challenge throughout is to be guided by the original text. One does not start out with a thesis or hypothesis that one tries to prove or disprove. Rather, phenomenological research is aimed at discovery and moving toward that goal of putting aside, as far as possible, one's own preconceptions.

In the dialogal method, the writing about their own experiences is a step that helps the researchers to identify or bracket their assumptions. What is most essential here, though, is that the group members develop a relationship of trust such that they feel free to identify and discuss the assumptions that are at work within the group. Each of us tends to be unaware of many of our own presuppositions, and it is often easier for others to see what we take for granted. Although it can be painful or embarrassing to have someone point out one's biases or preconceptions, this can also be a liberating experience in that it allows one to see what previously was hidden or obscure. Such an experience of insight can be very gratifying. Moreover, many of the assumptions that are identified are ones that the group members arrive at together.

On the basis of the above comparison, I would conclude that these two methods, while disparate in a number of ways, have in common an emphasis on basing analyses or interpretations on empirical evidence (i.e., descriptions) and on identifying and putting aside assumptions about the phenomena being studied so that it is possible to make genuine discoveries about these phenomena. It is also fair to say that they focus on the meaning of what is being said rather than on the words in a literal sense. In other words, while phenomenology is an approach that uses descriptions, it is not just a descriptive discipline that is content to summarize or repeat what various subjects or informants tell the researchers. This is an important issue because being faithful to what someone says is not the same as taking what they say literally. As we find all too often in our everyday lives, others may hear our words but fail to grasp their meaning.

Conclusion: The Question of Science

So let us for the last time look at the question of whether one should call phenomenological psychology scientific. Some of the connotations of what science involves seem to be that it is not scientific, if we define science primarily in terms of the models provided by biology and physics. Yet, it is clear, to me at least, that it can be a systematic and valid way to arrive at a meaningful and useful understanding of human beings' experience and behavior. Again, one's answer to this question depends both on what one means by science and on how one takes into account and conceives of the critical difference between studying nature and studying fellow human beings. Human beings, after all, look back at researchers in a way that rocks and plants do not. This is part of the reason that Wundt questioned the adequacy of experimental methods for studying complex mental processes. In a related vein, his contemporary, the German philosopher Wilhelm Dilthey (1833–1911), wrote extensively about the distinction between what he called studies of the mind or spirit (*Geisteswissenschaften:* geistes = spirit; *wissenschaften* = knowledge) and studies of nature (*Naturwissenschaften*), and how each of these domains required different approaches because of the difference in their subject matter.[61] The distinguished anthropologist Clifford Geertz has recently affirmed this point. While he expresses regret that the debate about what is scientific and what is not has generated more heat than light, he adds:

> But in one respect they have been useful. They have made it clear that using the term "science" to cover everything from string theory to psychoanalysis is not a happy idea, because doing so elides the difficult fact that the ways in which we try to understand and deal with the physical world and the ways in which we try to understand and deal with the social ones are not altogether the same. The methods of research, the aims of inquiry, and the standards of judgment all differ.[62]

Perhaps, part of the problem is that the attempt to determine what is scientific ends up being too abstract unless one places the issue within a particular context or relates it to a specific question. In an article presenting guidelines for reviewing qualitative research, three psychologists—Robert Elliott, Constance Fischer, and David Rennie—focus in on a fundamental issue when they suggest that "ultimately, the value of any scientific method must be evaluated in light of its ability to provide meaningful and useful answers to the questions that motivated the research in the first place."[63] The specific criteria they present for evaluating qualitative research include specifying the researchers' perspective and assumptions, making available adequate information about the research participants, using examples to indicate what the method involved and the kind of concrete data the conclusions were based on, providing some means for checking the categories or conclusions they arrive at, presenting their understanding in a coherent way, giving some indication of whether the findings are thought to apply generally or to rather specific instances, and finally, ensuring stimulating research that evokes resonance in readers. These criteria are ones that I believe most qualitative researchers, including phenomenologists, would agree with, at least in principle. So this means that it is possible to make at least some tentative evaluations of what constitutes good practice—and even good science—within the domain of phenomenological and qualitative research. I will not attempt to address the question of the relationship between phenomenological and qualitative research here, except to indicate that the former is one major tradition within the realm of qualitative research. (For an overview of types of qualitative research, see the excellent article by Lisa Tsoi Hoshmand in the *Counseling Psychologist.*[64])

The reality is that if you want to get the answer to certain kinds of questions, you have to use phenomenological or related methods. Consider the following: What is the relationship of physicians to their careers?[65] How do young children experience time?[66] What is women's experience of pain during childbirth?[67] What is it like for people who have had a heart attack?[68] What is the experience of psychiatric patients when they leave the hospital and return to the community?[69] How do the communications of deaf children differ from those of hearing children?[70] What is life like when one is aging?[71] These are all important questions, with significant practical implications for medical and psychiatric practice, for education, and for public policy. Answers to these questions, and many more, are provided by systematic studies of human experience.[72] These types of questions cannot be answered in any meaningful way by traditional psychological methods, which, by their very nature, presuppose the validity of human experience. At the same time, there are many university departments of psychology (as well as departments in related social sciences), journals, and funding agencies that do not support or acknowledge the legitimacy of phenomenological and other qualitative methods. They would argue that these methods are not scientific. However, it is not the case that these critics have carefully examined these methods and have found them wanting. The judgment is typically based on the assumption that because they do not resemble mainstream psychological methods, which in turn emulate those of the natural sciences, they are not scientific. This brings me back to the assertion of Elliott, Fischer, and Rennie that the value of any scientific method must be evaluated in light of its ability to provide meaningful and useful answers to the questions that motivated the research in the first place, as well as in light of its adherence to the notion that any method must be appropriate to the phenomenon being studied.

It is my hope that the contributions of phenomenological research will continue to grow and that they will gain increasing recognition for their value in deepening our understanding of what it means to be human. Holding a mirror up to human nature is a difficult and often messy business. But it is also immensely rewarding to participate in such research and there is much to be gained, I tried to demonstrate, from being exposed to it. I agree with Giorgi that psychology's emulation of the natural sciences, whatever its other merits, has obstructed the pursuit of studying human beings in human terms. No doubt the debate about the meaning of science, and the relationship between natural and human science, will continue for years to come. Yet I believe it is evident that phenomenological psychological studies do hold up a mirror to human nature and provide us with valid and meaningful insights into particular phenomena and situations. If this does not constitute human science, then what does?

CHAPTER 6

Interpersonal Relations and Transcendence

The familiar swallows up everything. It is bottomless. When experience fades into the familiar, it loses substance, it becomes a ghost.

Donnell Stern[1]

Whether we like it or not, the depths in us are always throwing up treasure.

John O'Donohue[2]

A s I was beginning this chapter, I looked through some of the descriptions of transformations in relationships that I had collected over the years. One in particular caught my attention. It is a story with the same theme as those presented in the first chapter, but like any deeply human and personal story, it also stands by itself. Because it was written by a former student whom I have gotten to know quite well over the years, it is especially meaningful to me. As I use it to discuss the connection between the interpersonal and transcendence, I will also focus on two other stories from earlier chapters, one from Chapter 2, on disillusionment, and one from Chapter 3, on forgiveness.

I start this chapter with these accounts because the term "transcendence" is often used to imply something otherworldly, religious, metaphysical, or outside of the lives of ordinary people and relationships. Thus, transcendence is commonly presented as a movement beyond the actual, the embodied, and the historical, something that should be studied by theologians or philosophers, but not by psychologists. In contrast, I argue that transcendence is a central thread in the fabric of ordinary human existence.

Later in this chapter I will discuss some of the meanings given to transcendence. But here I will briefly indicate how I use the term, if only to dispel the impression that it is an abstraction, remote from the actual. The

Oxford English Dictionary specifies that "to transcend" is to pass over or go beyond. This suggests that transcendence involves a venturing into, or an opening up to, something new. In his discussion of consciousness, Amedeo Giorgi has suggested that "the essence of consciousness is intentionality, not awareness. *Intentionality* [italics in original] means that consciousness always is directed to an object that transcends the act in which it appears. Basically, this means that consciousness is a principle of openness."[3] For the moment, then, let me suggest that transcendence is akin to openness, a movement toward the new, and is thus a key feature of our humanity.

If transcendence is a central thread in human existence, then why do we need to be reminded of it or to reflect on its meaning? Most basically, because its presence is all but denied in what we read and in how we talk about human life. The problem is not that the word is rarely used, but that what it points to is overlooked. Related words like "freedom" and, as I have suggested, "openness" do come up. If anything, freedom is spoken of all too much, especially in the last several years in the United States. The way in which freedom is discussed, especially in political discourse and in advertising (two arenas that are increasingly difficult to distinguish), tends to trivialize it. As consumers we are told that we have the freedom to choose from among a large number of options—"it is all about choice." Politicians remind us that we live in a free country where anyone can become president. The equation of freedom with consumer choice gives it a superficial meaning, and the claim that anyone can become president is misleading, to say the least.

Social scientists are generally on the other side of the fence—no breezy optimism or glib pronouncements about freedom, on their part. They rarely discuss openness or transcendence but instead focus on what appear to be the determining factors in our lives. Thirty years ago, B. F. Skinner, the radical behaviorist, argued that all behavior is controlled by environmental factors. As the title of his best selling book *Beyond Freedom and Dignity* suggests, Skinner believed that the notion of freedom was an illusion that, ironically, distracted us from the task of modifying the environment in our own self-interest.[4] However, while Skinner's behaviorism has been to a large extent displaced by psychological approaches that place greater emphasis on cognitive and cultural factors, this does not mean that determinism has gone by the board. For example, social constructionism, a relatively new approach within the social sciences, highlights how our behavior and experience are shaped by social and cultural factors, including language. The concepts used are different from those of Skinner, but the emphasis on how we are controlled is not.

Indeed, these factors are powerful determinants of human action and thought. But should we accept that we are locked inside the world view of our culture, or that the self is simply a reflection of one's culture and society, as some social constructionists, such as Kenneth Gergen, appear to imply?[5] Belief in this image of humans as blinded by their culture shows up in surprising places. For example, in his book *Hitler's Willing Executioners,* the historian Daniel Goldhagen insists that the Germans who killed Jews must be regarded as moral agents. But only a few pages later, he takes on a very different perspective. He writes, "During the Nazi period, and even long before, most Germans could no more emerge with cognitive models foreign to their society—with a certain aboriginal people's model of the mind, for example—than they could speak fluent Romanian without ever being exposed to it."[6] So here we have a clear statement of a deterministic position, and, again, one that captures an important truth. Yet it is also one that is misleading in its implication that cultures are monolithic and in its negation of persons as agents.

There is another reason why it is important to reflect on transcendence. As the psychoanalyst Donnell Stern (quoted earlier) says so aptly, "The familiar swallows up everything." The core dimensions of our existence become all but invisible. As the saying goes, fish do not notice the water in which they swim. Transcendence as familiar, and therefore hidden, is analogous to the love that exists between two people who have known each other for a long time. Love is evidence of our capacity for transcendence, for moving beyond ourselves to embrace and genuinely come to know another. The love that exists between two people is what animates the relationship, just as transcendence is a centrally defining quality of human existence. The particular shape and character of this love goes almost unnoticed, even as it is quietly expressed in everyday acts of kindness, expressions of interest, and meaningful touches and glances. At times, when the two lovers are at odds or engage in heated arguments, this love seems to disappear altogether or at least go into hiding. Then, when there is a crisis (e.g., one of the partners in the relationship becomes ill), or when there is an anniversary or another occasion to celebrate the relationship, this love, which is otherwise unobtrusive or concealed, moves into the foreground. Just as we can sometimes barely tolerate being loved or loving another, our capacity for transcendence is a mixed blessing. At times, it too is more than we can bear. I will say more about that later.

Three Stories

With this brief prelude, I turn to three stories from everyday life: a story of the growth of love in a family, a story of disillusionment, and a story of forgiveness. Each illustrates how, in a situation of crisis or change, transcendence becomes visible.

A Story of Love

Erica's story revolves around her changing relationship with her mother. She recalls the tension that existed between the two of them when she was in the eighth grade, a time when she began to rebel against her parents. Although she loved her parents, she also resented them because she was not sure they loved each other. Rather, it seemed to her that they were wrapped up in raising and looking after her sister and her, to the detriment of their own relationship. To make matters worse, the family had recently moved to a rural community, far away from Erica's friends and the world that was familiar to her. In this new place she was lonely and isolated. And it was her mother whom she held accountable for the move. But this was only one of her grievances. She believed that her mother was involved in her and her sister's lives to the point of excluding their father. Moreover, she saw her mother as holding herself up as "perfect" while being highly critical of her daughters and other people.

Yet, underneath Erica's anger, there was a deep desire for her mother to accept and love her and her father. Accordingly, she bought a special planter for her mother for Valentine's Day. When her mother did not respond as enthusiastically to this gift as Erica had wished, she got up during the night and broke the planter, leaving it for her mother to find in the morning. This was Erica's way of punishing her mother for not sufficiently appreciating her.

Then, when Erica's father was diagnosed with cancer, everything in the family changed. The parents and the children reached out to one another during this very difficult and painful time and quietly started to express more of their feelings. In Erica's own words: "For the first time I felt and saw love I could not see before." She saw her mother taking care of her dying husband day and night, and she saw a look on her mother's face that she had never seen before. In heartrending language, Erica describes how the change in her perception of her mother reached a new level when her father died: "The day of my father's funeral I heard my mother crying at his casket and she said something in such a way that my feelings about her changed forever. She said, 'Paul, I loved you always, I love you.' I finally saw her as a real woman who had lost her husband."

In the months and years that followed, a genuine friendship developed between mother and daughter. Erica was amazed at how her mother developed as a person, showing greater sensitivity to others as well as becoming more expressive and independent. She and her mother now communicated much more openly. One incident in particular illustrates how much their relationship had changed. Her mother consulted with

Erica when she had her first date several years after the death of husband, and Erica noticed that her mother was excited and nervous, as if she were a young girl again. In reflecting upon this incident, Erica writes, "When she shared these feelings with me, I saw her as a person who has the same kind of feelings as everyone else." She concludes her story with a statement of appreciation of her mother as "one of the most special and beautiful people that I know."

A Story of Disillusionment and Healing

This story is taken from Chapter 2; it was written by Jasmine when she was a college student. As you may recall, she tells how her father rescued her from drowning when she and her family were on a boat that capsized while they were escaping from a Southeast Asian country. Thinking back to this incident, which happened when she was six, she wrote, "My father was not only my hero, but he was the hero of several other people whom he saved." She continued to hear good things about her father from other family members and the larger community once the whole family had settled in the United States. And when her parents argued, as they often did, Jasmine blamed her mother for demanding too much of her father. When Jasmine heard that her father had left his family in order to live with his long-term mistress, she became very bitter: "I felt betrayed and I was angry at myself for admiring and respecting my father while all those years he was lying to all of us by pretending to be a just man." She wondered if he had ever really cared for his family. Like so many others, Jasmine came to believe that she had been fooled by the person she admired and concluded that she had been foolish in trusting someone so much. Later, her father returned home with a terminal illness. Jasmine did not cry when he died.

Two years after her father's death, Jasmine still had not discussed what had happened in her relationship with him with anyone. She wrote that this period was so incredibly painful for her that she became preoccupied with thoughts of ending her life, not surprisingly so, given that the father who had been the hero of her life was also the man who had betrayed her. Eventually, she went on a religious retreat and confided in the retreat director, a priest who reminded her of her father—"the loving, gentle man I always knew before my anger and hurt had overshadowed his goodness." This was the beginning of a process of healing, which gave her a sense of inner peace, enabled her to forgive her father, and thus allowed her the sense that he was still part of her life. In a real sense, her conversations with the priest gave her back her life.

A Story of Forgiveness

This is my own story, which has already appeared in Chapter 3. Once again, the following is the brief description that I wrote when my colleague Jan Rowe and I, along with four graduate students, started our project on forgiving another in the mid-1980s:

It was a summer evening, and I was agonizing over a romantic relationship that had reached a painful conclusion and left me feeling devalued, hurt, and angry. To relieve my distress, I decided to go for a walk in my neighborhood. The sun had not yet set and the colors of the trees and flowers were still vivid; in spite of being upset, I enjoyed their beauty as well as the peacefulness of the evening and the balmy air. I walked with no particular agenda in mind; what happened next was completely unexpected. The following is part of what I wrote about this event several years after the fact:

> My anger and hurt vanished as I was thinking about Heather, but this time as another human being who was struggling, and who basically did not mean me any harm. It is not accurate, I am realizing, to suggest that I just thought that; it was more like an image that emerged for me, an image that was not as much seen as felt. I felt healed; blame and anger vanished, and there was a larger dimension of this whole experience that I can only describe in religious language: a sense of transcendence, of the future opening up, of a sense of presence, not of a personal being, but of connecting to something larger than myself and yet still having an experience of myself as me.[7]

Reflections on Growth and Transcendence

These stories are about growth and change. We have seen how a relationship between a parent and a child can grow into much more of an adult-to-adult friendship, how one's faith in an idealized parent can be shattered, and how one can move, unexpectedly, from anger and demoralization to forgiveness and a sense of wholeness within oneself. Erica's story is also implicitly one of disillusionment that gives way to a more mature understanding of others. When she was in the eighth grade, she was very upset because she did not believe her parents loved each other in the way that she thought they should. Later, as a young adult, she realized that "the storybook family relationship that I wanted was only in my mind," indicating how her disappointment was replaced by an acceptance of her parents. She also moved from being critical of and blaming her mother for her own unhappiness to loving and appreciating her. In this story we see how both mother and daughter grow and become more open, developing an intimate relationship in the process. Jasmine found a way to reconcile with her father—albeit after his death—and I moved beyond anger to an acceptance of Heather as a fellow human being.

These stories deal with particular relationships while also illustrating the basic human capacity for taking into account more of the reality of the world around us. They point to how we are continually reminded of how the people in our lives exceed, or are at odds with, the images we have of them. Erica, for example, reached out to members of her family, took into account their points of view, and arrived at a new self-understanding. She matured in the process of facing the very painful reality that her father was dying and acknowledging that her mother was changing right in front of her eyes. Human beings are capable of learning, changing, growing, transforming—this is what I mean by saying that transcending (moving beyond) is at the core of human existence. Transcendence is an ongoing process that we do not usually reflect on—we are continually moving forward in time, toward the new as well as to specific experiences where we are quite conscious of having opened up to something or someone because of what has happened in a dramatic or unexpected way.

The reality that Erica and her family confronted—the inevitability of death—allowed for a deepening of their relationship. You may recall that at the end of Chapter 1, I referred to Jan H. van den Berg's assertion that it is the light of death that makes those whom we love dear to us. It would be just as true to say that in facing the reality of death, whether our own or that of someone we care about, we awaken more fully to the fact that we are alive. In a similar vein, the philosopher Martin Heidegger describes death as a basic horizon of human existence. The personal acceptance of our mortality, he argues, can move us toward an authentic existence, that is, acknowledging that this is my life and that I am choosing to live it in the way that I do.[8] In Erica's story we see that as her mother becomes much more expressive of her love for her father as he moves toward death, Erica becomes more open and engaged with her family and, in that sense, becomes more alive.

Awareness of mortality is also a key issue in an essay written by the psychiatrist Donald Cohen, but for him, as for van den Berg, love and death are to be considered together. In reflecting upon his life and work, he refers to the philosopher Franz Rosenzweig's book *The Star of Redemption*. Cohen writes, "It begins with a stark statement that I used to believe as a matter of faith,"[9] namely, that the beginning of philosophy is the fear of death. Over time, however, Cohen becomes convinced that this is not the whole story. "I have learned from our research and clinical work, and in my own bones, that philosophy can be fed from a sweeter well, a more caring source—from Eros and its earthly representative, our first and continually replenished capacities for loving."[10] By way of conclusion, Cohen summarizes a tenet of Rosenzweig's philosophy with which he is in full agreement: "The engagement with others in the fullness of relationship is the foundation which sustains us in facing life and enduring suffering, the motivation for development from birth until the very end."[11] This statement likely brings to mind different experiences for each one of us—the support received from a loving grandparent, the importance of friendship during especially difficult times, the compassion of colleagues during a time of loss, or the kindness of neighbors during a time of illness. Unfortunately, the notion of the power of love has become a cliché. But by no means does this change the fact that the kind of love Erica describes and Cohen refers to is a sustaining force in our lives, in contrast to other forces, such as hate and brutality, that tear us apart. Thus, love, as sustaining and forward-moving, is closely related to transcendence. Or perhaps, as I suggested earlier, it would be better to say that genuine love is a form of transcendence.

Here we come back to transcendence, this perplexing concept that has multiple meanings. In my view, two of the meanings are fundamental. The first meaning is the one to which I have already referred, namely, our forward-moving nature, or capacity for openness. That is, humans are *transcending* beings, however much we also get stuck and go out of our way to deny reality. The second meaning refers not to us but to that or who is beyond us, in other words, *the transcendent*. Philosophers speak of the transcendent as the "truly other."[12] The term *truly other* is used to refer to God, or the ground of our being, as the Lutheran theologian Paul Tillich would say. What is it that distinguishes a genuine experience of the divine (or whatever word that might be most appropriate) from other experiences of the world around us?

One way of formulating the difference succinctly is as follows: "I do not lay hold of the transcendent. It lays hold of me."[13] This formulation has an intuitive appeal. Those who describe mystical or spiritual experiences often speak of being overwhelmed, left speechless, and being shaken to the core of their being. Some philosophers also use the term "truly other" to refer to our experience of another person when it exceeds our image of him or her, along the lines of Emmanuel Levinas's notion of the infinity of the face of the Other.[14] In this kind of experience, there is an overturning of our ordinary stance toward the other. It is not that I look at the other and make sense of who this person is. Rather, the other reveals him or herself to me as I shift from "constructing the other" to being open or receptive to him or her. In Chapter 1, I wrote of the themes of surprise and wonder and the separateness of the other as two key themes in the experience of seeing the other "as if for the first time."

The philosophers who explore this second meaning of transcendence question what it means to experience something that is beyond our grasp (and even beyond our comprehension), be it another person or God. They ask whether it is contradictory to say that we experience something as beyond us because the fact of our experiencing it seems to imply it is not truly beyond us. Moreover, how can we even begin to describe that which transcends us and is beyond our grasp or our concepts? These are certainly important and difficult questions. But however eloquently and thoughtfully philosophers explore the meaning of transcendence, they rarely turn to the concrete lives of human beings to inform their reflections. For example, the contributors to James Faulconer's book *Transcendence in Philosophy and Religion,* from which I have drawn extensively in this discussion, approach the problems and questions around this issue on a completely theoretical basis.[15] This is especially striking since most of these philosophers are phenomenological in orientation. One of them even writes, "If phenomenology can, better than most philosophies, do justice to transcendence, this is because the only authority it recognizes is experience."[16] But, ironically, the only manner in which experience shows up in this book is through analyses of texts that discuss the *concept* of experience.

So what do we find when we examine accounts of perceiving the other person as truly other? As I have already indicated, we find that people speak of being surprised and having a sense of wonder as the other becomes present to them. As we saw in Chapter 1, when people describe seeing another person as truly other—seeing him or her as if for the first time—they describe the other as an understandable fellow human being, as someone whose point of view becomes accessible to them. But the other is both truly other *and* someone one can empathize with. This is certainly the case in my description of forgiving Heather. So it seems to me that the philosophers who use the term "truly other" to refer both to the experience of another person and to that of the divine or God are doing so prematurely. That there is some difference is evident, also, in the fact that it is easier to find words to describe interpersonal than religious or spiritual experiences. In my description of forgiveness, I had few words to refer to what I called larger dimensions of the experience—"a sense of transcendence, of the future opening up, of a sense of presence, not of a personal being, but of connecting with something larger than myself." I was awestruck and at a loss for words.

It might be more on target to look at some of our experiences of nature as examples of encountering the "truly other." Ron Silvers, a friend and colleague of mine in Toronto, is, among other things, an Arctic photographer. At one time, he showed me enlargements of pictures that he had taken of icebergs in the Arctic. Just to see pictures of these enormous creations takes one's breath away. Icebergs are beautiful, majestic, and truly awesome. But it is not just their scale and majesty that strike awe in us. It is also the realization that they are inanimate and "indifferent" to us and that they exist in a time frame outside of human history even while also overlapping with it. We have a similar experience of awe when we look at the stars. They are light years away from us and utterly beyond our grasp and our comprehension. Thus, there is no question that they transcend us even as we perceive them; they bring home to us just how small we are in relation to the cosmos.

But whether we are discussing our experience of the other as other or as nature, or of the mystery beyond words that surrounds us, we are at least implicitly dealing with situations in which we are aware of our smallness, of our limitation, and our vulnerability. This kind of radical openness is not one that we come to readily or often. In fact, it is fair to say that we go to great lengths to avoid being open to others and to the fundamental realities of our existence. We do not have to look far to see evidence of human opacity and the refusal to acknowledge the humanity of others. How, then, can one make sense out of the fact that while we have the capacity for transcendence, we seem to make so little use of it? And perhaps, more strikingly, what are we to make of the fact that although human beings have the capacity to affirm one another and become intimate, our daily news confronts us with the pervasiveness of human destructiveness, both at the personal and at the collective level. Spouses, most typically men, kill their partners and, at times, their children as well. Terrorists kill innocent civilians as do the armed forces that fight terrorism. These are not intellectual puzzles but deeply disturbing and frightening realities, even for those of us who are not directly affected. While humans are able to be open and to love, human history offers us a tragic and disturbing portrait of who we are all too much of the time: creatures who are blind and hateful. In what follows, I outline the position of Ernest Becker, who spent his life addressing these apparently irreconcilable sides of human nature.

Transcendence, Death Awareness, and Evil

Becker, a cultural anthropologist who in his all too brief life (1924–1974) defied the conventions of his discipline by tackling fundamental questions about human life (rather than studying human behavior in minute detail or constructing esoteric theories), was one of the rare social scientists who affirmed humanity's capacity for transcendence and yet also was keenly aware of the extent to which "stuckness" describes much of human behavior and history. Like the behaviorists and the social constructionists whom I mentioned earlier in this chapter, Becker emphasizes the environmental and social factors that shape our behavior even while his view of human nature is more multidimensional than is theirs.

From his perspective it is obvious that imagination and thought allow us to see beyond our immediate circumstances and that such a seeing beyond might especially occur during a time of crisis, as was true for Erica and her family. However, while Becker would not have disagreed with van den Berg's assertion that our awareness of mortality deepens our appreciation of life and of those we love, he was convinced that this awareness also brings with it much darker consequence. He believed that being open to the reality of our existence, specifically to our vulnerability, helplessness, and relative insignificance, is more than we can ordinarily tolerate. Looking at the stars, we do indeed register something of our smallness and the transient nature of our lives, but it is an awareness that we work hard to subdue. In his view, "the idea of death, the fear of it, haunts the human animal like nothing else; it is a mainspring of human activity—activity designed largely to avoid the fatality of death, to overcome it by denying in some way that it is the final destiny of man [sic]."[17] We are acutely aware of our bodily nature, even while we are also embarrassed and even baffled by it.

But one might reasonably ask: how is it possible to deny something that is undeniable? As the saying goes, two things in life are certain—death and taxes. And yet, as Freud (to whom Becker is indebted) points out, the human capacity for evading reality is extraordinary. As it turns out, the forms of evasion amount to something more subtle than a direct denial of the fact that one will die physically. Becker argues that our capacity for transcendence (for taking into account circumstances and events beyond those with which we are immediately faced) leads us, collectively and individually, to resort to denial and obfuscation. Or, to put this in plainer language: on one side we have a capacity for openness and yet, on the other side, this gives rise to a variety of forms of defensiveness that reduce our consciousness of reality. The most elementary and disturbing realities that we are capable of taking into account are the very realities that we deny or hide from ourselves.

In his book *Escape from Evil*, Becker suggests that one of the functions of culture (which includes mythology, religion, politics, and nationalism) is to give us a sense of immortality and significance:

Everything cultural is fabricated and given meaning by the mind, a meaning that was not given by physical nature. Culture is in this sense "supernatural," and all systematizations of culture have in the end the same goal: to raise men [sic] above nature, to assure them that in some ways their lives count in the universe more than physical things count.[18]

In other words, as human beings, we have collectively created meaning structures and ideologies that allow us to feel as if we are part of a larger whole and that assure us we will survive death, either literally or metaphorically. For example, Christianity promises the believer eternal life—resurrection of the body and life in heaven. Political leaders believe that they will have a place in history and are concerned, especially toward the end of their term in office, with how they will be remembered. Those who have children and grandchildren hope that they will be remembered after their death and take comfort in the continued survival of their family name, along with their genes. The wealthy can donate money to institutions and have buildings, conference rooms, or endowed chairs named after them. Similarly, architects assume that the buildings they have designed will survive as monuments to their creativity. Authors hope that their words will continue to be read after they are long gone. The terrorists who hijacked the four planes on 9/11 believed that they would be remembered as martyrs and would be rewarded with all sorts of delights in the next world.

Joining groups and holding on to belief systems provide us with a sense of power and significance exceeding that which we have in our individual existence. Members of fan clubs for the famous or the infamous may feel as if the fame of the persons they idolize rubs off on them, giving their own lives more excitement and importance. In Chapter 2, we saw how idealizing another, be it a father or cousin or anyone else, gives the person who idealizes a sense of special significance. It is uplifting to remind oneself that this person who is really capable, powerful, or attractive really loves one, supports one, or cares for one. Patriotism or political party affiliation also provides people with a sense of being part of something larger than oneself and thus creates the illusion of leaving behind one's own vulnerable and—in the larger scheme of things—not so terribly significant existence. We become deeply attached to the communities (be they religious, familial, nationalistic, or professional) to which we belong. Thus, we do not take kindly to those who disparage our heroes, whether they are religious figures, athletes, movie stars, or political leaders.

Following the death of former president Ronald Reagan, there was a nationwide commemoration of the achievements of this man who, to a number of Americans, was larger than life. Others were quick to point to failures in his record as president as well as to his personal limitations. In turn, Reagan's admirers wrote indignant letters to the editor, suggesting it was unseemly to criticize someone who was no longer able to defend himself and for whom thousands of people still grieved. Implicitly, they were saying: "If you criticize the one whom we admire, you are diminishing us." As Becker asserts, "Each person nourishes his immortality in the ideology of self-perpetuation to which he gives his allegiance; this gives his life the only abiding significance it can have. No wonder men go into rage over fine points of belief; if your adversary wins the argument about truth, you die."[19]

Of course, the dying of which Becker speaks here is symbolic in that it involves the collapse of our meaning structure rather than physical annihilation. But threats to our core beliefs are deeply disturbing and are not always so easily distinguished from threats to our physical existence. Think of the shock and sense of betrayal that Jasmine experienced when her father, who had saved her life and was her hero, turned out to have abandoned his family to be with his mistress. She was so distressed that she had thoughts of ending her life and was so ashamed of being let down by her father that she did not discuss this deep loss and the resulting despair with anyone.

Our attempts to deny our mortality and our vulnerability through our participation in collective cultural ideologies are deeply rooted in our lives. This does not change the fact that, Becker points out, that these attempts are only partially successful. They are flawed even though we internalize these ideologies or structures of practices and beliefs at our mother's breast, so to speak, and consequently many of them are largely unconscious or taken for granted. We do not choose belief systems the way we choose cars or clothes (although these purchases may be more intimately connected to our core beliefs than we care to admit), nor can we readily discard them at will. They live within us, at an embodied level, as is evidenced by our strong physical and emotional reactions when they are affirmed or brought into question. Nonetheless, these symbolic systems are flawed, precarious, and even destructive. As Becker says so powerfully, "The terror of death still rumbles underneath the cultural repression."[20] No wonder we often have murderous thoughts toward the messenger who brings us disturbing news.

The first flaw or problem is that no matter how much we deny or repress, these defense mechanisms are always subject to being undermined as we are continually reminded of our mortality and finitude. This is particularly true, Becker suggests, in the postmodern age where long-standing religious and political traditions are continually being questioned. Few would disagree with this assertion. For example, the eminent student of world religions Huston Smith has written, "If anything characterizes 'modernity,' it is a loss of faith in transcendence, in a reality that encompasses but surpasses our quotidian affairs."[21] As we saw in the previous chapter, even our faith in the objectivity and certainty of science has been seriously challenged.

Second, our aggressively defensive posture (I am a true believer, you are an infidel) leads to all kinds of conflicts with people whose persuasions are different from our own, as well as with those within our own circle of believers who express doubts or start to define their lives in new ways. Becker's basic theory of evil is that it arises from our compulsion to treat those whose ways of life and belief we perceive as fundamentally discordant with our own as less than human, thus giving rise to individual or collective "crusades" against infidels, heretics, or evildoers. Or, as the cultural critic Terry Eagleton writes, "Immortality and immorality are closely allied."[22] Thus, the "solution" to our fear of death generates enormous problems.

Third, by subscribing to and embodying these collectively and individually constructed ideologies (Becker also refers to them as heroic illusions because they claim to lift us out of the ordinary and mask off our fear of death), we narrow the range of our experiencing, thinking, and behaving. We cut ourselves off from some of our own impulses and possibilities and from meaningful contact with a variety of people, and of course, we also avoid some of our own fears and anxieties. This form of escaping from oneself obviously comes at a price. As Becker puts it, "We fashion unfreedom as a bribe for self-perpetuation."[23]

Erica, in a small way, had developed her own ideology as to how life should be lived and what love should look like. She was critical of her parents, and especially of her mother, for not living up to her ideal of showing love to each other. Her mother, too, seemed to have lived a restricted life, perhaps with clear definitions as to how one should behave. Similarly, Jasmine initially had a highly idealized image of her father as the powerful protector.

I have discussed Becker's views not because I agree with everything he says, but because I regard his analysis as distinctive and insightful. More than anyone else I know, he helps to explain the link between our extraordinary human capacity for thinking and imagining and our capacity for creating trouble for other humans and for ourselves. He shows how our capacity for transcendence is a mixed blessing. On the one hand, our potential openness to the world and others is truly remarkable, as we have already seen. For example, forgiveness frees us from the stranglehold of the past as is evident in my description. On the other hand, Becker argues strongly that our attempts to hide and overcome our sense of insignificance and mortality have terrible consequences. As Rollo May has stated, "Violence has its breeding ground in impotence and apathy."[24]

There is empirical evidence from recent social psychological studies that provide solid support for Becker's theory about the nature of human violence. A group of social psychologists, most notably Jeff Greenberg, Tom Pyszczynski, and Sheldon Solomon, has been drawing upon Becker's writings to develop what they describe as "terror management theory."[25] The research that has been carried out by this group and other social psychologists across the globe is very extensive. Here I will just summarize the gist of some of these experiments. Researchers have consistently found that if you remind research participants of their mortality, even in a subtle way, they will typically more strongly identify with their own group and beliefs and will react more negatively to those who represent opposing beliefs or values. For example, municipal court judges who read a questionnaire about death set much higher bonds for women charged with prostitution than did another group of judges who were not exposed to such a questionnaire.[26]

Of course, no one author can adequately account for a topic as fundamental and complex as human evil. I would encourage anyone who is interested in understanding evil from a different perspective to read James Waller's fine but chilling book *Becoming Evil: How Ordinary People Commit Genocide and Mass Killings.*[27] Waller focuses on the historical and psychological factors that contribute to particular episodes of human destructiveness and thereby complements Becker's more general approach.

For Becker, the term "transcendence" has closely related and yet different meanings. First, he uses it to refer to the fact that we can imagine and anticipate our own death, and thus take into account much more than our immediate circumstances. We are not, to use an old psychological phrase, stimulus bound. As a variation of this meaning, he refers to self-transcendence, by which he is referring to the multitude of ways in which we can escape awareness of our own mortality and vulnerability by participating in a larger framework of meaning, such as that provided by religious tradition. However, while Becker is clear that this "self-transcendence" is often problematic, he also acknowledges that human beings cannot help but seek a larger significance and meaning in their lives. Yet he also points out that there are many different forms of self-transcendence. Some are more life-giving, open, and less destructive than others. These include our intimate relationships with others. As Daniel Liechty puts it, "Such relationships are authentic and valued intimations of transcendence."[28]

The second meaning of transcendence in Becker's writings is more closely related to the religious or spiritual realm. Here he is speaking of that which transcends us. From reading Becker, one might reach the conclusion that his attitude toward religions is primarily negative and critical—after all, they are a source of "heroic illusions" and crusades against those who are construed to be evil. Glenn Hughes, a philosopher who has written about humanity's encounter with transcendent or ultimate meaning throughout history, is critical of Becker in this regard. Hughes chastises Becker for failing to break with the assumption of modern social science that any experience of a transcendent being (i.e., God) must be a projection or invention of the human psyche.[29] In contrast to Hughes, I think it is fair to say that Becker's position on this issue is ambiguous, at least in his writings. During an interview with Sam Keen that took place while he was dying from cancer, Becker stated his personal position plainly. "I think the birth of my first child was the miracle that woke me up to the idea of God more than anything else, seeing something pop in from the void and seeing how significant it was, unexpected, and how much beyond our powers, and our ken." A moment earlier, he had spoken of the importance of recognizing that "beyond the absurdity of one's life, beyond the apparent injustice of things, beyond the human viewpoint, beyond what is happening to us, there is the fact of the tremendous creative energies of the cosmos which are using us for some purposes we don't know."[30] Here transcendence refers to our experience of something that vastly exceeds us, such as what he describes as the "creative energies of the universe," or the world of nature, in the face of which we feel so small and insignificant.

Certainly, Becker is not the only writer who has something to say about transcendence. With this in mind, I will explore how this concept is understood in selected theological, philosophical, and psychological literature, and will conclude with an explanation of what I mean by speaking of transcendence as a key but overlooked constituent of human life. I undertake this exploration because I believe it is of critical importance that we think carefully about what it means to be human and that our understanding of our humanity be both realistic and congruent with the breadth and range of our experience. At the beginning of Chapter 4 ("Experiencing the Humanity of the Disturbed Person"), I mentioned my concern that reductionistic or one-dimensional theories about human behavior hold too much sway or are applied rather literally to human beings in general as well as to the mentally ill. By focusing on transcendence, I seek to broaden our view of who we are.

Meanings and Misunderstandings of Transcendence

However, to broaden our view of human nature does not mean that we should romanticize or exaggerate human possibility or freedom. To begin with, we ought to be suspicious, or at least skeptical, when someone claims to have transcended a problem or a situation. Speaking of transcendence, in many cases, amounts to little more than denial, dressed up in fancier and superficially more spiritual or philosophical garb. In everyday parlance, we say, "That's all behind me," or "I am over that." There are people who all too soon speak of having forgiven someone for having injured them; they have not even had time to fully appreciate how much they have been hurt, let alone allowing themselves to get angry about what has happened. Certainly, all of us want to have "closure" and leave unpleasantness and pain behind, but life is typically a lot messier than that.

This tendency to claim that one has left difficult or painful events behind is not just a characteristic of individual persons, but is applicable to larger communities as well. Lance Morrow has suggested that it is particularly true of this country. In his gripping book *Evil: An Investigation,* he writes, "The American idea is to leave the damage behind, and not look back; every evil will be transcended as the nation goes on, reinventing itself in better and more prosperous ways."[31] In spite of slavery, the decimation of Native Americans, the active participation in the overthrow of democratically elected governments such as Chile and Guatemala, an unusually brutal prison system, and the neglect of children's welfare, many Americans think of evil as something "out there." We would rather not admit that it is a distinctly human reality that we, like all other people in the world, have to confront both individually and collectively. Thus, I want to emphasize that when I am using the word "transcendence," I am not endorsing a "new age" philosophy that anything is possible or minimizing the harsh realities of human life. As Gabriel Marcel points out (Chapter 1), there is a vast difference between optimism (the ideology that everything will turn out well) and hopefulness (an attitude of openness to possibility).

We must acknowledge that our lives are shaped and defined by many factors. The structure and form of the human body are something from which we cannot escape. Although genetics does not equal destiny, it would be foolish to underestimate the role hereditary factors play in determining our physical nature as well as our personality and abilities. We are also social and cultural beings to the core, as the social constructionists rightly point out. That is, we are born into families, and we learn a particular language within a particular community at a certain point in history. There is no question that we are deeply influenced by our social and physical circumstances. Waller's book on how ordinary people become killers outlines in painful detail how readily individuals can be persuaded and coerced into brutalizing and murdering fellow human beings.

As I indicated at the beginning of this chapter, social constructionists argue that individuals are largely a reflection of the society in which they live. In this interpretation of humanity, our thoughts, outlook, and behavior are determined by social processes and expectations and thus everyone is locked within his or her culture's world view. Many social constructionists also hold that social scientists are equally unable to see beyond the theory, methods, and language of their own professional traditions. It is taken for granted that none of us, either as persons or scientists, can break through such parochial barriers to reach some larger truth. In this context, Hughes is correct in asserting that the social sciences appear to deny the possibility that humans can have a genuine experience of the transcendent.

Nonetheless, we must acknowledge that those who subscribe to social constructionism have something important to say. When we travel or learn a new language or go through a significant change in circumstances, we quickly realize that how we previously saw the world was conditioned by language, locale, and other aspects of our circumstance. And yet our susceptibility to social and institutional pressures is often minimized or exploited by those who hold political and economic power. Advertising that is aimed at millions of people everywhere calls on us to exercise our individuality by buying a mass produced product, and political parties loudly proclaim that they stand for greater freedom even while they attempt to get us to embrace ideologies that require us to become thoughtless and thereby less free.

The irony is that if we uncritically subscribe to the belief that individual free will is a primary determinant of what happens in our lives, we reduce our own freedom. The dramatic increase in obesity in North America is a case in point. One interpretation of the problem is that people need to make better choices, that is walk more and eat less fast food. In short, they need to exercise their free will. However, to focus primarily on individual choices is to overlook the obvious, namely, that when there is a widespread problem, the factors contributing to it are also widespread and beyond individual control. With the rising rates of obesity, there is general agreement as to what these factors are: people eat out more often and cook less, and many people eat at fast food restaurants that serve high calorie meals; people get less and less exercise, and this is increasingly true for children who spend much of their time in front of computers or televisions. It is simply implausible that individual choices will change this situation, because the environment in which so many people live supports weight gain and poor health. What are more likely to work are collective efforts to change the environment by bringing pressure on fast-food chains to have healthier offerings and to improve the quality of food available to students in their schools, for example. A report by the Institute of Medicine (a division of the National Academy of Sciences) recommends a series of measures, including the regulation of TV commercials aimed at children, more physical education in schools, and changes in the kinds of foods that are available in schools.[32] This institute is calling for structural changes rather than appealing to individual responsibility, much to the dismay of a number of major corporations selling the products, such as snacks and fast foods, that are associated with the increase in obesity.

Unfortunately, the discussion around the issue of freedom (another way of speaking about transcendence) and determinism is typically presented in such a way as to preclude the development of a deeper understanding. Typically, determinism and freedom are conceptualized as mutually exclusive positions. But there is another way of thinking of this problem, one that recognizes the merit and the limitations of each position. As Allen Wheelis writes so eloquently:

> We must affirm freedom and responsibility without denying that we are the product of circumstance, and must affirm that we are the product of circumstance without denying that we have the freedom to transcend that causality to become something which could not even have been previsioned from the circumstances which shaped us.[33]

In a similar vein, Maurice Merleau-Ponty points out that we are both subjects and objects—we act and we are acted upon, we shape and we are shaped. Our freedom is limited or situated.[34] To lose sight of our freedom, however limited it is, would be tragic, just as it would be foolish to lose sight of how our behavior is influenced by the context in which it occurs. However, as my colleague George Kunz points out, there is little appreciation for paradox or ambiguity either in common sense or in psychology, however much these concepts resonate with what we find to be the case on a daily basis.[35]

We need to find the evidence of our capacity for transcendence, for freedom and choice, within the context of the everyday, and not in some special "sphere" of life. As with the love between two people, transcendence is woven into the fabric of ordinary existence and thus only occasionally becomes perceptible. The alternative view is that the transcendent and the everyday (or the immanent) exists as two alternative realms. One of the writers who holds to this position is Richard Cox, who, as both minister and psychologist, is concerned that "psychology has become an idol of this age" and has unduly influenced churches.[36] He argues that psychology deals with the immanent, that which is within our reach and that, broadly speaking, falls within the category of sensory experience. From psychology, psychotherapists learn about human behavior and specific techniques for intervention. In contrast, religion deals with that which is beyond us, the transcendent. Psychotherapists who hope to bring genuine hope to their clients must themselves become transcending and thus they must draw upon religious tradition. In essence, Cox attempts to reverse the relationship between psychology and religion. Rather than religion looking to psychology for its own enhancement, psychology—in the form of psychotherapy—needs to look to religion.

The problem with Cox's argument is not that he seeks to distinguish religion from psychology, that he sees psychology as being focused on the immanent, or that he sees the transcendent (i.e., God or the divine) as the special province of religion. The problem, instead, is that he implicitly divides existence into two realms, the immanent and the transcendent, with some people (and therapists) who have reached the level of becoming transcending and some who have not. In this way, he defines transcendence as an admirable attainment rather than as a basic human capacity, as "the ability to rise above the norm, the belief that humans can achieve a state of mind and resultant life style that is grounded in the extraordinary, the 'yet not seen but believed.'"[37] And yet we have seen descriptions from several young people, including Mary (in Chapter 1) and Erica, who have "transcended" or moved beyond their previous understanding of life and others to gain much greater empathy and appreciation for those close to them. None of these descriptions, we should note, make any obvious reference to a transcendent or divine reality.

Although Cox's assertion that psychology deals with the immanent is reasonable, it is not entirely accurate. There are several psychologists who have written specifically about transcendence. Most notable among them is the humanistic psychologist Abraham Maslow. Maslow is famous for his work on the hierarchy of needs, self-actualization, and peak experiences. In one of his early discussions of peak experiences, he describes these unforgettable and positive experiences as similar to mystical experiences. Those who had these experiences, he writes, reported that they had the "feeling that they had really seen the ultimate truth, the essence of things, the secret of life, as if veils had been pulled aside."[38] At first he believed that these peak experiences were confined to very healthy or self-actualized people, but this turned out not to be the case. He found that they occurred in ordinary people as well. Peak experiences seemed to be almost religious in nature insofar as they included a sense of profound peace and made a connection with a profound reality. And yet, the context for them was not typically religious in any obvious way. They occurred in nature, in lovemaking, in moments of creativity, and in listening to or playing music. In his early article, Maslow does not use the word transcendence, but seven years later he writes about "the various meanings of transcendence." While he lists no less than thirty-five meanings, they all refer to what he calls a higher level of human consciousness.[39] Each of the meanings involves a moving beyond, be it beyond one's own cultural preconceptions, personal regret, objective time, fear about death, self-preoccupation, the opinions of one's peers, or separation of self from others. To put it differently and more positively, each meaning refers to a giving of oneself to something beyond oneself, the personal, and the cultural, such as in reaching out to another, feeling connected with people from another period in history or another country, or having a mystical experience.

In a related psychological study of transcendence, Jenny Wade interviewed ninety people who had a spiritual or religious experience during lovemaking. In her book *Transcendent Sex,* Wade refers to three features of her respondents' experiences that define what she means by this term. First, they involve an altered state, such as experiencing radiant light surrounding oneself or penetrating one's being.[40] These events are completely unexpected. Second, the experiences are attributed to a supernatural force or spirit, even by those who regarded themselves as agnostic or atheists. Third, the experiences arise in the context of the relationship between the two lovers. Oddly enough, although the study has much in common with Maslow's research on peak experiences, there are no references at all to his work in Wade's book.

In both Maslow's and Wade's studies, the emphasis is primarily on transcendent experiences, on the kinds of unusual experiences that people have. But they also refer to the fact that humans are capable of having such experiences, that is, they also refer (at least in passing) to persons as transcending. This is the point to which I want to return.

Merleau-Ponty has written clearly about this issue. First, he locates transcendence within the person and defines it in terms of his or her activity. Thus, "Consciousness is transcendence through and through, not transcendence undergone . . . but active transcendence."[41] He further describes the forward-moving nature of human actions "as the violent transition from what I have to what I aim to have, from what I am to what I intend to be."[42] His description of human agency and striving brings to mind issues from earlier chapters. Disillusionment, as we saw with Jasmine, can be seen as the disruption of one's taken-for-granted forward movement, the blocking of one's intentions and the collapse of one's life project. In many cases, we are not very much aware of our life direction, our values, and our hopes until we run into obstacles that highlight what we really care about. Remember how Benjamin, the man whose girlfriend left him for another man, spoke of having been "betrayed by a dream" (Chapter 2)? This woman was someone he idealized; she was pretty, athletic, and well educated, and he enjoyed her company. She represented the "good life," and his relationship with her allowed him to move forward. His hopefulness gave way to despair. Yet, Benjamin was determined not be embittered by this experience, even while unsure whether this was possible. It took him a year of searching and seeking help and guidance from others before he felt that he was again on solid ground.

Merleau-Ponty's view of transcendence as active dovetails with that of Heidegger. In *The Basic Problems of Phenomenology*, Heidegger affirms that we are actively transcendent and relational beings. Thus, our existence is, from the outset, an existence with others, and others "join with us in constituting the world."[43] All of the three stories on which I have focused in this chapter show how the world is constituted with others. Jasmine's world collapsed when her father left the family, but she was healed through her relationship with the priest who reminded her of her father. Erica's world was dramatically transformed as she joined with her mother and sister in grieving the loss of her father and gradually established a deep friendship with her mother. I found myself uprooted and distressed after my relationship with Heather ended, but found peace, at least for a time, after my experience of forgiving her. We are able, Heidegger writes, to experience the other as a thou and can similarly be experienced as a "thou" by the other. "For 'thou' means 'you who are with me in a world.'"[44] In other words, as we saw in Chapter 1, we are able to experience the agency and subjectivity—the transcendence of the other.

In further elaborating on transcendence, Heidegger argues that it has everything to do with our relationship to time. His view is that humans embody time (or temporality) and that it is not a force outside of us. That is, the concepts of past, present, and future always imply a being who has a history, a being who remembers, anticipates, and acts. He rejects the more popular use of the term where the transcendent refers to God, the otherworldly, or that which is "outside of the subject." The notion of that which is outside the subject implies that the person is somehow encapsulated within the self rather than existing in, and open to, the world. Heidegger concludes that "the transcending beings are not the objects—things can never transcend or be transcendent; rather it is the 'subjects'—in the rightly understood sense of the *Dasein*—which transcend, step through and over themselves."[45] This is similar to Merleau-Ponty's position, in that existence or selfhood always entails stepping beyond. And this is where Heidegger makes the link with temporality—to be human is to be constantly living with possibility and the emergence of new meaning.

The notion of living with possibility and experiencing oneself as agent plays a prominent role in the thought of Karl Rahner. Rahner was a highly creative Catholic theologian who had studied with Heidegger and whose work fits with the spirit of phenomenology. He insisted that in scholarly disciplines that deal with human existence, including theology, assertions must be tested against personal experience of the issue at hand,[46] and that "reflection never totally includes the original experience."[47] In other words, there is always more to our experience than we can conceptualize or put into words.

When Rahner refers to transcendence, he speaks of "transcendental experience," by which he means not a specific experience but an aspect of our existence in the world. In any experience, we are aware not just of what we experience but also, at least implicitly, of ourselves as experiencing and knowing subjects. Secondly, in agreement with Wheelis and Merleau-Ponty, Rahner reminds us that the self that we are is not just determined by the surrounding world. This awareness brings with it at least some degree of realization that we have responsibility and freedom to choose. Moreover, we are aware of ourselves as having knowledge and also as recognizing the limits of this knowledge, as well as of the inexhaustible possibilities of finding out more.

It is worth noting, however, that to speak of transcendence as active (as do Heidegger, Merleau-Ponty, and Rahner) is not the complete story. Previously, I used the word openness to refer to transcendence, and a number of the transforming experiences described in this book appear to have more to do with a movement that occurs as one *allows* oneself to be affected or moved by an event or by another. Many of those who spoke of their experience of forgiving another (see Chapter 3) used expressions such as "being freed" or "released from a burden" when referring to the change they experienced. Additionally, our experience of ourselves varies depending upon our circumstances. At times, we have a particularly striking and thematic awareness of ourselves as experiencing and knowing subjects; at other times, we feel unfree and buffeted by the events and forces around us. Again, this is evident in the three stories. At the beginning of each one, the narrator describes himself or herself as caught, hopeless, and personally diminished. Erica lives with a keen sense of disappointment and is convinced that unless her parents relate differently to each other and to her, she will continue to feel lonely and unloved. Jasmine writes of how devastated she is after being disillusioned by her father; she is not sure she wants to go on living. I describe myself as feeling devalued, hurt, and angry, and, in addition, having no idea how to move beyond these feelings. By the end of his or her story, it is fair to say, each person describes feeling free and moving forward with his or her life.

Concluding Questions

I have suggested that the word "transcendence" has two basic meanings. First, that it refers to human beings' fundamental capacity for moving forward and for opening up to the new and taking it into account. Following

Becker's analysis, we can also see that this power is a two-edged sword. We cannot help but turn away in some measure from the realities that we see, more or less dimly, that overwhelm and frighten us. Our basic situation is one of mortality and vulnerability. The other meaning of transcendence refers to that to which we are open, namely, that which exceeds us or is beyond us: the truly other. An obvious question emerges here: what is the relationship between these two meanings of transcendence? This is a question to which there is no ready answer, and it is not a question that psychologists are well qualified to address. Hence, I turn to James Jones, both a psychoanalyst and a religious thinker, who addresses the question of whether transcendence belongs to the subject or whether the transcendent is outside of the subject. His conclusion is that this is a false dichotomy. Jones writes, "The experience of the sacred has a transcendental, numinous quality not because the sacred is wholly other but because such experience resonates with the primal depths of selfhood."[48] If nothing else, this quote gives us something to think about.

In the next chapter, I will discuss contemporary psychological theories of the interpersonal and of the self to see to what extent there is a recognition, even if only implicitly, of the dimension of transcendence in human life. Earlier in this chapter, I referred to Hughes's assertion that the social sciences deny the possibility of human beings having a genuine experience of the transcendent. Whether or not this criticism is entirely justified, we might wonder if psychology leaves room for the possibility of experiencing another person as genuinely other and for opening up to the new. Second, we might ask, what should we turn to in view of psychology's failure to adequately take into account this dimension of our humanity?

CHAPTER 7

Psychology, Transcendence, and Everyday Life

Philosophy does not raise questions and does not provide answers that would little by little fill in the blanks. The questions are within our life, within our history.

Maurice Merleau-Ponty[1]

We are most creative and sense other possibilities that transcend our ordinary experience when we leave ourselves behind.

Karen Armstrong[2]

In this chapter, I will look at selected approaches in contemporary psychology to follow up on the question raised at the end of the previous chapter. That is, is there some way in which psychology recognizes the dimension of transcendence and openness in human life, and especially in relationships? Do these approaches address the arena of interpersonal relationships in a way that resonates with the kinds of phenomena that this book has addressed? Or, to put it a bit differently, what kind of guide is contemporary psychology when it comes to helping us to understand and appreciate the depth of our relationships? And given that psychology is necessarily limited, where else is one to look, how else is one to proceed? I end this chapter (and this book) with reflections on these basic questions.

We have already seen in Chapter 2 that there is a good deal of material, mainly from psychoanalytic writers, that addresses disillusionment in a meaningful away. In contrast, in Chapter 3, we found that although there is a growing body of psychological research on forgiveness, it gives little attention to the study of forgiveness as it is experienced in everyday life. In fact, this research largely ignores features of the forgiveness experience that challenge mainstream psychological images of human nature. In what follows I will turn to two other approaches within psychology. First, there is the study of closeness and intimacy, a relatively new focus within the broader field of what is called "relationship science." Second, within psychoanalysis, there has been a growing emphasis within the last twenty-five years on understanding human behavior from an interpersonal point of view. This point of view is concerned with looking at behavior and experience in relational terms. Classical psychoanalysis, in contrast, focused on what might be happening in the mind of the person considered as a separate individual.[3]

To study these areas of psychology is rather like entering a subculture or worldview with its own language, history, and set of assumptions. It is necessary to take these assumptions into account, as I will demonstrate in what follows, even though my treatment of these approaches will be both brief and selective. With these introductory comments in place, let me turn to the area of intimacy and closeness study.

Intimacy and Closeness Studies

In Karen Prager's well-known book *The Psychology of Intimacy,* an especially telling statement appears on the very first page: "If any reason needs to be given for devoting an entire book to intimacy, it is that intimacy is good for people."[4] She argues that people who have close relationships have better health, are less likely to be overwhelmed by stress, and have more adaptive lives than those who do not have such relationships. This quick summary of psychological research is a reminder that we are social beings, depending on others for companionship, support, putting things in perspective, enjoyment, and so on, and not just for help with everyday tasks, such as holding the ladder steady when we climb up on a roof.

What I find noteworthy about the statement is the implication that writing a scholarly book on intimacy needs justification. Psychologists write about other areas, such as psychopathology or memory, without providing any justification for what they are doing. And isn't intimacy, after all, at the very heart of what psychologists want to address? The answer is both "yes" and "no." The "yes" has to do with the importance of intimacy and closeness for human beings. The "no" has to do with the difficulty of bringing an approach that uses scientific research methods, narrowly construed, to bear on a dimension of human existence that is subtle, complex, and resists objectification. Consider the introduction to a recent publication, the *Handbook of Closeness and Intimacy* (2004), edited by Debra Mashek and Arthur Aron, which is similarly telling. It states that "this *Handbook* establishes closeness and intimacy as a substantial sub area of relationship *science* [my emphasis] as well as reflects the latest thinking of a large group of top researchers in the rapidly advancing field of relationship science."[5] They describe the development of this field in the last twenty years, including the advent of conferences, journals, and research grants. "Young scientists," they tell us, "are building their laboratories and focusing their attention on these topics," "there is now a solid and rapidly growing body of relationship knowledge," and this field is commanding increasing respect.[6]

Two things are going on here. First, the editors of this book are trying to convince us that this is an important book, and of course, this is what editors are supposed to do. But what is much more interesting is that they are also arguing for the scientific respectability of this endeavor. This is suggested by their references to laboratories, relationship science, and a growing body of knowledge. And there are no less than three chapters in this book that address the question of how closeness and intimacy can be measured. Further, for those of us in academia, there is an additional and related message between the lines: the research of young faculty who work in this area should be taken as seriously as research carried in any other area of psychology, for example, when these researchers apply for tenure. Intimacy and closeness may be "soft" topics or more broadly, as Prager suggests, any concept that comes from everyday life has "fuzzy boundaries."[7] Nonetheless, we are assured such topics can be studied using standard, well-respected scientific methods and that research carried out in this way will advance our understanding of relationships. One social psychology textbook claims, rather simplistically, that a field is scientific if it uses scientific procedures.[8] However, as was discussed in Chapter 5, this assumes that the questions are settled regarding how one defines science and by what criteria one determines which procedures are scientific, an assumption that is very much open to question. We also learned that phenomenologists, among others, have insisted that what makes a research method scientific is that it is suitable for studying a certain type of experience. In other words, which method is *scientific* ought to be determined by the context in which it is used rather than on the basis of whether the method has been deemed appropriate in disciplines other than psychology.

Fortunately, this field of study is by no means monolithic either in terms of methods or perspectives. In addition, there are a number of researchers and theorists who believe that one needs to take seriously the experience of ordinary persons. For example, Steve Duck, a prominent scholar in this field, has written that research is useful insofar as its findings connect with how people understand and think about their own relationships. He goes on to repudiate the notion that the ways in which scientists make sense of human interactions are somehow objective whereas ordinary persons' sense making is of dubious value.[9] About forty years earlier, Fritz Heider developed an argument similar to Duck's in his classic study *The Psychology of Interpersonal Relations*. Heider made a successful effort to build a scientific theory based upon a systematic examination of people's "naïve" psychology of daily inter-action. It would be foolish, he suggests to his "hard-nosed" colleagues, to ignore this "naïve" psychology. First, it guides people's conduct, and second, it has developed over hundred of years of human history.[10]

Duck is especially interested in how relationships change, and believes that making sense of change requires attention to the contexts in which people interact. In his discussion of change, he makes a critical distinction relevant to the type of interpersonal epiphany considered in Chapter 1. He states that one can know another from the point of view of an observer who is aware of specific facts about this person. This is different from understanding another where "I organize my knowledge of you in a way that includes knowledge of how you make sense of things (i.e., viewing you from the inside). The switch from knowing someone to understanding the person is an important one in relationships."[11] In a similar vein, Aron, Mashek, and Aron argue that closeness involves *inclusion*. To be close to someone includes "experiencing (consciously or unconsciously) the world to some extent from the other's point of view."[12] Neither Duck nor these researchers raise the question of how it is possible for us to take in another's point of view, one that is beyond our own (that which I have referred to as our capacity for transcendence), but they do acknowledge that such a shift takes place.

What Aron and her colleagues do emphasize is the pragmatic value of relationships, a value that Prager refers to in the introduction to her book. They assert that we are motivated to "include the other in the self" because in so doing we increase our own resources both in a material sense (e.g., "I can borrow my friend's car") and a psychological sense (e.g., through my closeness to the other I learn about new ways of thinking about problems and thereby grow as a person). One of the major and early theories in the area of close relationships, developed by the social psychologists John Thibaut and Harold Kelley, is called the *social exchange theory*. Also focused on the pragmatic, this model assumes that we approach relationships in a way similar to how we approach business exchanges. If the partners in a relationship gain from their association, it will continue, but whenever one partner feels like he or she is giving too much and getting too little (as determined by to his or her level of expectations), the relationship is likely to be terminated by that person.[13]

Looking at relationships through the lens of loss and gain has merit. No doubt many of us think of relationships in these terms, at least at times. Yet it is also obvious that such a lens is both limited and limiting. Its basic concepts are imported from the disciplines of economics and business, arenas where concepts such as loyalty and fidelity are not central and where the notion of the discovery of a new reality makes little sense. When we look back at some of the stories in Chapter 1, it becomes evident that this economic model does not, and cannot, do justice to much of what happens between people. Think, for example, of Mary's recognition of what it was like for her mother to have been home alone for years with no one but small children for company. Or, think of how Vanessa came to see her younger brother in a much more empathic way when she attended his wrestling match. These experiences involved a fundamental shift in the ground of both of these relationships.

As this discussion suggests, theory plays a prominent role in social psychological and relationship research. From out of their own theoretical perspectives, a number of researchers believe that conversation plays a primary role in the development of relationships. Duck holds to this position. Self-disclosure, that is, telling someone else about personal aspects of one's own life and experience, is widely seen as one form of talk that is particularly significant for deepening relationships.[14] There is no reason to dispute that talk and self-disclosure are significant in relationships and in intimacy. The closeness that developed between Sheila and Sr. Lois (Chapter 1) came about as Sheila spoke more frankly both about her relationship with her boyfriend and about the conflict with her teacher and as Sr. Lois listened and was genuinely accepting of what she was told. This is certainly also congruent with what a number of researchers have found.[15] However, some of the stories in Chapter 1 present a different scenario. As you may recall, in a number of cases, it was not self-disclosure but the opportunity to see the other person differently because of his or her powerful response to a particular event that created an increased sense of closeness. Along this line, Lisa Register and Tracy Henley found in their phenomenological study of people's experience of intimacy that a number of their respondents emphasized the nonverbal dimension of their intimate moments. The respondents had difficulty expressing in words what had happened, because it was so surprising and profound. Moreover, speaking was not necessarily a central aspect of the interaction, but exchange of glances or touch did play a major role. Altogether, their study suggested that intimate moments are transformative and transcendent.[16]

One of the more interesting chapters in the *Handbook of Closeness and Intimacy* was written by Prager and Linda Roberts and deals with intimate connections in couples' relationships. What makes this chapter especially interesting is that the authors draw upon Prager's clinical experience working with couples and that they consider in their discussion the relationship between two characters in E. M. Forster's novel *Howard's End*. This relationship vividly exemplifies how strong the obstacles to intimacy may be. A number of the themes that Prager and Roberts raise are very similar to what we learned about close contact in Chapter 1. They suggest that intimacy requires "access to a true and authentic self"[17] and refer to a number of psychotherapists (as opposed to researchers) who support that contention. I would add that the kinds of experiences described in Chapter 1 suggest that connecting with another simultaneously involves connecting with the self. Intimate contact, they further suggest, involves complete attention to, as well as positive regard for, the other and is characterized by immediacy. Yes, indeed.

This quick overview of intimacy and closeness studies suggests that the very nature of the phenomenon studied does, in a sense, push back and reveal itself even when the research methods and theories have a restrictive influence. But researchers obviously know more than what their research allows them to conclude, especially when it comes to an area as close to all of us as intimacy.

The Interpersonal Direction in Psychoanalysis

As Lewis Aron has pointed out, a broad range of contemporary psychoanalysts have firmly acknowledged that the psychoanalytic process must be understood in interactive and interpersonal terms.[18] That is, the person of the analyst and the person of the patient need to be seen in terms of their mutual influence. And, of course, if this applies to psychoanalysis, then it applies to all kinds of relationship. One might even say, perhaps a bit uncharitably, that psychoanalysis has caught up with Martin Buber, Emmanuel Levinas, Gabriel Marcel, and other philosophers who have been insisting for years that human existence is, at its core, interpersonal.

In what follows I discuss primarily the approach developed by George Atwood, Robert Stolorow, and their associates. Their work, which was discussed briefly in Chapter 5, powerfully exemplifies the movement of psychoanalysis toward paying closer attention to experience. And the problems of their approach are as instructive as their strengths.

In their 1979 book, Atwood and Stolorow set out their ambitious agenda. First of all, they want to develop a "metapsychology free framework to guide clinical psychoanalytic conceptualization and treatment."[19] By a metapsychology-free framework, they mean a framework that is based on clinical observations and does not contain unwarranted assumptions, be they derived from philosophy or the natural sciences, about the nature of persons and about relationships. Sigmund Freud, for example, assumed that the basic human motivations are instinctual in nature and that people had an inner mental world that is separate from an outer world of objective reality. These assumptions, Atwood and Stolorow would assert, were not based directly on what Freud learned from his treatment of patients but on his medical training and the worldview of his times. In a recent publication, Stolorow, Atwood, and Donna Orange identify a number of problematic assumptions, implicit in psychoanalytic theorizing (including Freud's) that they describe as Cartesian, that is, as derived from the seventeenth century French philosopher René Descartes.[20] These include the belief that a person is an isolated entity; that there is a split between mind and body, self and other, inner and outer; and that it is possible to arrive at certainty about the nature of reality. Instead, these reformers argue, personality psychologists ought to return to the basic task of "understanding the experience and conducts of persons,"[21] a phrase that shows up consistently in their subsequent writings. In their 1979 book *Faces in a Cloud*, they call their approach "psychoanalytic phenomenology." The phenomenological emphasis refers to their effort to develop principles for understanding human beings in ways that stay close to the actual experience of patient and therapist. For them, the in-depth case study is the best method for returning to the study of experience. I would note that although Atwood and Stolorow used the term "phenomenology" often, their discussion of phenomenological philosophers, such as Edmund Husserl and Martin Heidegger, is at best cursory, and they appear to have scant knowledge of developments in phenomenological psychology subsequent to the late 1950s.

By using the term psychoanalytic, they state their intention to look not just at immediate experience but also at how patients' developmental history has affected their view of self and others, and how their behavior is guided by these beliefs.[22]

What do Atwood, Stolorow, and their associates believe is at stake here? They share the conviction, held by phenomenological psychologists, that psychology should study the world of human experience. Part of their motivation is based on their clinical experience: if psychoanalysts are guided by wrong assumptions, then problems will arise that undercut their attempt to be therapeutic. The example they give, in many of their publications, runs something like this: a psychoanalyst who follows a classical Freudian model is working with a seriously disturbed patient who has suffered trauma and losses growing up. As a result the patient is very vulnerable and has a fragmented and chaotic way of relating to the world that is very different from the way the therapist and other reasonably mature adults relate to the world. The therapist is unable to empathize with what the patient says, because of the person's disturbance, but fails to realize this.

As a consequence of the therapist's inability to understand what is going on for the patient, the patient becomes even more disturbed. However, the therapist who thinks of himself and the patient as two separate individuals (in the fashion of the Cartesian idea of a split between self and other, which is part of the Freudian metapsychology) attributes the increased disturbance to the patient's pathology, thus creating further estrangement between himself and the patient.

This example illustrates how metapsychological assumptions have a direct and negative effect on the effectiveness of psychotherapy. If one were to follow Atwood and Stolorow's approach, one would be attuned to the way in which behavior should be understood in a relational context and, as a result, would reflect on the effect of one's own behavior on the other person. The recognition of the interpersonal dimension of psychology and psychiatry has a long history, going back to the work of Harry Stack Sullivan,[23] but for Atwood and Stolorow it becomes foundational. Given their emphasis, they soon rename their perspective an *intersubjective* approach to psychoanalysis. In their 1987 work, they state that "the concept of an intersubjective field gradually crystallized in our thinking as the central explanatory construct for guiding psychoanalytic theory, research, and treatment."[24] This is akin to Heidegger's notion of being-in-the-world, which emphasizes that all of human existence is relational to the core.[25]

There is a second aspect of their agenda that follows from the first. One of the problems with the study of personality, they argue, is that the field is fragmented because each theorist approaches the arena with his or her own ideological or conceptual images of human nature and reality. Hence the subtitle of their 1979 book: *Subjectivity in Personality Theory.* These images necessarily reflect the theorists' own personalities and their historical and cultural context. The worldviews of Carol Rogers, Freud, and Carl Jung, for instance, are quite distinctive, and no amount of data collection will resolve the differences among them.

So is there a solution to this fragmentation? Atwood and Stolorow believe that there is and that they can provide such a solution:

> If the science of human personality is ever to achieve a greater degree of consensus and generality, it must turn back on itself and question its own psychological foundation. There must be sustained study not only of phenomena that have always been its province, but also of the biasing subjective factors that contribute to its continuing diversity and fragmentation. Progress toward clarifying these predisposing influences can be achieved by a psycho-biographical method that systematically interprets the metapsychological ideas of personality theories in light of the critical formative experiences in the respective theorists' lives.[26]

Simply put, Atwood and Stolorow say that one should look at the theorists' perspectives and consider how their own background led them to bring into their work assumptions that had to do with their own lives (and life crises) rather than just the subject matter. The task is to separate out the *meta* from the *psychology* and, in this process, get closer to the experiential or phenomenological validity of these theories. Atwood and Stolorow anticipate an obvious objection to this project, namely, that they themselves cannot help but bring their own limited and biographically biased perspective to this task. Although they admit this is true, they also argue that they draw upon a broad range of psychological theories in looking at the biases of theorists, that they are continually looking at how theory relate to experienced realities, and that they bring to this task familiarity with the role of philosophical assumptions, such as those of Descartes. In response to a critique by George Frank of what he calls the "intersubjective school of psychoanalysis,"[27] Stolorow insists that he and his colleagues have not tried to create another school. "Rather, the intersubjective perspective offers a unifying framework for conceptualizing psychoanalytic work of all theoretical schools."[28]

The intersubjective perspective has made an important and positive contribution to psychoanalysis and to our understanding of human relationships. Atwood and Stolorow have emphasized the importance of moving away from doctrinaire presupposition and paying closer attention to the actual experience of therapist and patient. For example, they interpret "acting out" (e.g., through self-punitive acts or apparently random aggression toward others) as attempts to shore up a precarious sense of identity. (This is in contrast to the rather mechanical Freudian conception that "acting out" is the result of the failure of the ego to repress instinctual drives.[29]) There is the sense in a number of their case studies that their understanding of their patients emerges through the therapeutic dialogue. And no one could dispute the value—and even the critical importance—of their insistence that the patient's behavior must be understood as a function of his or her relationship to the therapist. The point, I would hope, is not that one should now blame therapists whereas previously therapists might have been inclined to blame patients. The more constructive implication of the emphasis on the intersubjective context is that one ought to attempt to understand both participants' behavior and experience in relational terms.

What are the limitations of this approach from the point of view of the emphasis on intimacy and transcendence? The first and inescapable limitation comes from the fact that the intersubjective approach, along with other psychoanalytic perspectives, is primarily based on clinical case studies. It is not just that the focus is on people with psychiatric problems but that the case study itself is really not all that descriptive. First, case studies are written from the perspective of the psychotherapist and, second, most of these cases do not tell you very much about the ongoing interactions between analyst and patient. We are presented, as the summary of Atwood and Stolorow's case in Chapter 4 indicates, with an overview that emphasizes the overall principles that the authors want to highlight. Thus, it is hard to know how "experience-near" their interpretations are. In fact, it is difficult to find any descriptions that are faithful to the experience of both therapist and client. One notable exception is a book coauthored by Irving Yalom (the therapist) and Ginny Elkin (the client), which includes their notes written after each session.[30] There is also the pioneering work of the humanistic psychologist Carl Rogers who audio-recorded therapy sessions, thus allowing others to examine for themselves the data on which he based his conclusions.[31]

When we take a close look at some key concepts in *Worlds of Experience*, a 2002 book by Atwood, Stolorow, and Orange, other problems become evident. They emphasize that in their view all experience is situated in a relational context. Then they add that "an intersubjective field—any system constituted by interacting experiential worlds—is neither a mode of experiencing nor a sharing of experience. It is the precondition for having any experience at all."[32] The words that, in my mind, seem odd are *system* and *experiential worlds*. What happened to the emphasis on the study of the experience and conduct of *persons?* And to belabor the obvious, worlds do not constitute, persons do.

An earlier article by Stolorow helps us to make sense of the reference to systems.[33] In it, he refers to the dynamic systems theory developed by Esther Thelen and Linda Smith[34] and suggests that it is "a source of powerful new metaphors for psychoanalysis."[35] So where does the dynamic systems theory come from? The answer is chemistry, physics, and mathematics. Moreover, it is an approach that is applied to a variety of realms from cloud formation to the behavior of children. Again, what happened to the experience of persons and the effort to keep metaphysical assumptions, even if in the guise of metaphors, out of psychology?

There is little explicit reference to the dynamic systems theory in *Worlds of Experience* (although Thelen and Smith's publications show up in the References). Yet this theory is nonetheless very much in evidence. For example, the authors state that "in dynamic intersubjective systems, the outcomes of developmental or therapeutic processes are emergent and unforecastable rather than preprogrammed or prescribable."[36] Interestingly, Stolorow seems to acknowledge that there are problems with the language of systems theory, stating that his viewpoint could be described as a "no-person psychology."[37]

From my perspective, the problem with the intersubjective systems approach of Stolorow and his colleagues becomes most evident when one looks at their argument with the psychoanalyst Jessica Benjamin around the issue of mutual recognition within psychotherapy, as well as human relations more generally. One of the most erudite thinkers among contemporary psychoanalysts, the late Stephen Mitchell, has described Benjamin's writings as creative and provocative. He notes that she draws upon the thought of the philosopher Georg Hegel and of a broad range of psychoanalytic thinkers, as well as feminist theory.[38] One of the arguments that Benjamin consistently makes is that any adequate theory of intersubjectivity must include the possibility of mutual recognition, that is, of each person seeing the other as someone who is a subject separate from the self.[39] She also makes plain her belief that the theories of Heinz Kohut and Atwood and Stolorow are lacking in this respect. In 1995 she presented her own view of the intersubjective as follows:

> Intersubjective theory postulates that the other must be recognized as another subject in order for the self to fully experience his or her subjectivity in the other's presence. This means that we have a need for recognition and that we have a capacity to recognize others in return, thus making mutual recognition possible.[40]

On the previous page, Benjamin had written, "What difference does the other make, the other who is truly perceived as outside, distinct from our mental field of operation?"[41] Her answer is that there is an obvious difference between the experience of seeing the other as other (to use Levinas's language) and seeing the other primarily as an "object" within one's own world. I should add that part of the context for her discussion of this distinction comes from her hope that children (and especially boys) increasingly learn to recognize their mothers as fellow subjects rather than remaining stuck in the notion that their mothers (and, by implication, other women) exist primarily to serve them.

Stolorow, Atwood, and Orange charge Benjamin with having reverted to Cartesian dualism insofar as she describes the other as "outside, distinct from our mental field of operations." They also express concern that the psychotherapist might start demanding the patient move toward recognizing him or her as a separate person or that this becomes a therapeutic goal.[42]

I see both of these criticisms as missing the point. In my estimation, Benjamin is too sophisticated an analyst to treat recognition as a moral imperative for her patients. Nor is it plausible, in the overall context of her writing, that "distinct from our metal field of operations" implies that she construes persons as "isolated minds" in any Cartesian sense. Rather, it seems to me that Benjamin's position is close to that of Buber and his emphasis on the growth of persons in the context of relationship of an I-Thou relationship. Moreover, the recognition of which she speaks closely resembles the stories that I presented in Chapter 1. Buber writes of the unique capacity of humans to create both distance and relationship: "Man [*sic*], as man, sets man at a distance and makes him independent; he lets the life of men like himself go on around him, and so, and he alone, is able to enter into relation in his own individual status, with those like himself."[43] The implications of Buber's position about our paradoxical human capacity is that we do not have to choose between a Cartesian dualism that conceives of persons as isolated monads and an intersubjective dynamic systems theory where the reality of the face-to-face relationship is not clearly acknowledged. In the moments of recognition described throughout this book, participation in the perspective of the other and awareness of his or separateness go together, just as awakening of the self and awakening to the other go together.

Conclusion: Where Else to Look, How Else to Proceed

We have seen how psychological theories and research illuminate aspects of relationships even while they leave some dimensions of these relationships in the shadows. It is probably not fair to assert, as Glenn Hughes does, that the social sciences, or at least psychology, altogether overlook or deny the transcendent dimensions of life. If we read between the lines in this literature, we can see glimpses of something beyond the limited images of the person to which many psychologists subscribe. Even if Atwood and Stolorow, for example, do not take us fully into the depth of human relationships, they do nonetheless take us quite a way in that direction. In any case, it is not as if we have to choose between being appreciative and being discerning, whether we are considering research and theories or people we know. It is neither prudent nor charitable to minimize the limitations of scholars or friends anymore than it is right or fair to minimize or overlook their strengths.

Where else do we turn, to whom do we turn for guidance, given the limitations of psychology, at least as I have presented them here? There are certainly writers in psychology and related disciplines who have a more complete vision of human life than the ones I have mentioned in this chapter. For example, in *Rediscovery of Awe,* the psychologist Kirk Schneider encourages us to look to mystery and wonderment as we seek transformation of our lives. The theologian Karl Rahner (discussed in

Chapter 6) explores the relationship between love of neighbor and love of God, while Buddhist writer Sharon Salzberg eloquently outlines an experiential understanding of faith. As we have already seen, Buber writes poetically of the spiritual dimension of intimate human relationships in his classics study *I and Thou*.[44] Reading can broaden our horizons and help us to find words and wisdom that enable us to look more deeply at our lives. But, as we know, reading can only take us so far; it can only affirm or support the path on which we are already walking or respond to the questions with which we are already struggling.

So where else do we look? This depends, of course, on who we are and what our circumstances are. Thus, the more powerful question is *how* should we look? To attempt to answer this second question, I will draw upon three stories that tell us something about the possibilities for discovery and hope in the midst of everyday life.

When I started this manuscript, some years ago, I did not know how I would end it, and, until a few months ago, I still did not know. That I now know how to bring this book to a conclusion is not owing to my own wisdom, knowledge, or research. It is due to something entirely different and unexpected. Let me explain.

I live in what is called an intentional community. My housemates and I share a large house owned by a nonprofit organization. One of the purposes of this organization is to make it possible for people with different backgrounds and income levels to live together in small self-governing groups. The youngest member of our community is a one-year-old boy named Milo. One Saturday afternoon, as I sat reading the paper in the kitchen, one of my housemates was also there playing with Milo, whose parents were doing their yoga exercises in the living room. At a certain moment, this housemate had to attend to something and she handed Milo to me. It was a clear winter day and so I held Milo up before the window so that he could look out on our backyard. To my surprise he did not squirm or become impatient. Instead, he was very attentive, just occasionally turning his head as if he wanted to take in all of what was out there, beyond the window. Something was holding his attention and I started to wonder what it might be.

With that question in mind, I started to look more closely at what was there in our backyard. At first, I concluded that there wasn't much of anything. There were no people, nor were there any squirrels running up and down the trees. So what was Milo looking at? I saw that the branches of the trees were moving in the wind, something that I had at first overlooked. I was reminded that for a young child the world is a wondrous place. Everything is new and amazing, and what we as adults have come to regard as ordinary is noteworthy, absorbing, and surprising. As I kept looking, I started to notice the interplay of shade and light, movement and stillness of each tree, with its own distinctive shape and character, and of the richness of this quiet and unspectacular, yet inspiring, view. This heightened sense of awareness of the world remained with me for the rest of the day. I felt as if I had been given a gift.

Recently, a friend told me of an extended vacation during which she had spent hours each day just watching the clouds move across the sky. When she described this experience, I was reminded of that afternoon. I thought about that brief time when Milo and I looked out the window and experienced the wonder of the world together, even while our perspectives were years apart.

There are many clichés about looking at something or someone with new eyes, just as there are clichés about what happens when we fall in love. But it is when something actually happens to us or when we make a discovery that these sayings come to life; they are no longer stale or trite. The phrase "My eyes were opened," used by one of the people whose story I included in Chapter 1, has an almost universal resonance. In my moments with Milo my eyes were opened. But to what? At one level the answer is, perhaps, to not much of anything. I had looked into our backyard thousands of times during the years I have lived in this house, and while I had appreciated the view, no one viewing had added up to an event that I remembered as such. What, then, was different here? This time Milo had been my guide, helping me to slow down and take a second look, allowing me to listen to the silence whispering in the trees and reverberating within me.

The simple truth is that we need help to see and experience the world more openly and more deeply, but it is not necessarily "experts" who can help us here. In our daily lives we are overwhelmed with sensations, with busyness and input, as has been extensively documented by a number of thoughtful social commentators, such as the Canadian writer Heather Menzies. In her studies of the cultural and economic change brought about by technology, Menzies helps us to see how the frantic pace of our individual lives reflects changes in our society and in our institutions.[45] In a similar vein, Thomas de Zengotita, an editor and cultural anthropologist, has written about what he calls the "Numbing of the American Mind." He asserts that in the contemporary world of media and communication technology, "Our minds are the product of total immersion in a daily experience saturated with fabrications to a degree unprecedented in human history. People have never had to cope with so much stuff, so many choices. In kind and number."[46] We find life, or at least the feeling of being alive, in busyness and in being constantly subject to sensations and the sensational. Yet this feeling is often numbness in disguise. We are touched on our skin and our minds, but not our soul. The feeling of realness, Zengotita argues, requires that we become attentive and take in the breeze and the stillness. But it is indeed often hard for us to dwell with something so apparently subtle and elusive.

It is even harder for us to be still and attentive when we are in the presence of another person. One measure of a comfortable friendship is that we can be silent together. Likewise, one measure of a seminar that is going well is that the participants can tolerate (or even welcome) pauses, moments of silence when something that has been said is allowed to register, when no one feels the need to rush in and fill up the space.

As Zengotita suggests, much of modern life works against our paying attention and being present. A change in circumstances, as we saw in Chapter 1, may provide us with the opportunity to move forward and to be surprised as we discover the new. But circumstance alone does not make such a discovery possible—we need, so to speak, to lean into them. Consider the following story, written by the Canadian religion writer Tom Harpur. He had set out to interview Mother Teresa in Calcutta, India, and to learn about the humanitarian work in which she and her coworkers were engaged. Not long after arriving in the city, he went to visit a hostel for the poor and the dying. To his dismay, the nurse accompanying him handed him some food and instructed him to feed a feverish Hindu man lying on a cot. This is not something that Harpur had bargained for; the man looked anything but appealing, and there was also the risk of contracting a fever. His description of what happened next is brief but memorable:

> Yet as I knelt to break the bread, put it in his mouth, and pour the broth when his eyes showed that he wanted to drink, a remarkable change took place in me. It was in no way anything to boast of—after all, the nurse really gave me no choice—but I found myself overcoming my revulsion and fear. I caught a glimpse of what Mother Teresa calls "the Christ in very man or woman" and was deeply moved.[47]

Although we do not know exactly what Harpur experienced, it is clear that this was for him an unforgettable and wondrous experience of another human being as in some way a transcendent being.

Of course, the point here is not that we must go to India in order to have an epiphany, anymore than we require a particular young child to show us the way. Seeing more and seeing differently is always a possibility, it is always right around the corner, or right in front of us. We just do not know the possibility is there until it materializes. And we often overlook that it materializes only with our help.

Let me conclude by going back, for a moment, to that apparently simple story (in Chapter 1) of a young woman who came to see her boyfriend's brother in a new light. As you may recall, Rachel was visiting Wayne's apartment and was looking in his refrigerator for something for the two of them to eat, when the phone rang and Wayne answered it. Rachel watched him as he spoke on the phone. "The light from the kitchen fell on him for a second and in that second I saw Wayne not as Gary's brother but as Wayne. The light falling on his face shadowed his eyes and brought out his cheekbones and his nose and sculptured his face. For that second I seemed to see what he was and what he could become."

This story does not involve self-disclosure, at least not in the sense that psychologists who study closeness understand it, where one person intentionally reveals something very personal to another. Yet, it is certainly an experience of intimacy, even if just from Rachel's perspective. It is a moment where she sees more of the fullness of Wayne's humanity and is touched by what she sees. Obviously she sees Wayne from her own point of view and yet she is witness to something of Wayne's relationship with his own world. It is, as I have argued throughout this book, evidence from everyday life that we are transcending beings who are capable of taking in that which transcends us, such as in this case another person who can look back at us. To be a person is to live in the world with others. And anytime we become truly present to this reality, we are both enriched and humbled. As we have seen, in such a moment we experience deep empathy, appreciation, or love for the other. It is also a moment when we, paradoxically, come to our senses as we allow ourselves to move past self-absorption and self-consciousness to a connectedness with something or someone that includes us and also surpasses our own boundaries. In discovering the other we rediscover our own capacity for openness, and it is this openness that is at the core of what it means to be a person.

Notes

Introduction

1. Maurice Natanson, "Anonymity and Recognition: Toward an Ontology of Social Roles," in *Condito Humana,* ed. W. Van Bayer and R. Griffith (Berlin: Springer Verlag, 1966), 32.
2. Salley Vickers, *The Other Side of You* (New York: Farrar, Straus and Giroux, 2006), 91.
3. George Steiner, *Real Presences* (Chicago: University of Chicago Press, 1991), 139.
4. Anita Shreve, *The Pilot's Wife* (Boston: Little Brown, 1998).
5. cf. Edmund Husserl, *The Crisis of European Sciences and Transcendental Phenomenology,* trans. David Karr (Evanston, IL: Northwestern University Press, 1970).
6. Walter Lowrie, *A Short Life of Kierkegaard* (Princeton, NJ: Princeton University Press, 1970), 115. The existential tradition in philosophy was so named because it was concerned with addressing the dilemmas and problems of personal existence. In the years roughly subsequent to the 1930s, the existential and phenomenological traditions converged.
7. The phenomenological philosopher Maurice Merleau-Ponty strongly emphasized the necessity of an ongoing movement between the flow of experience and reflection on experience, in the *Phenomenology of Perception,* trans. Colin Smith (New York: Routledge and Kegan Paul, 1962) and especially in his last manuscript, published as *The Visible and the Invisible,* trans. Alphonso Lingis (Evanston, IL: Northwestern University Press, 1968). For a helpful interpretation of the latter text, see John Sallis, *Phenomenology and the Return to Beginnings* (Pittsburgh: Duquesne University Press, 1973).
8. G. K. Chesterton, "The Blast of the Book," in *The Scandal of Father Brown* (New York: Dodd, Mead & Company, 1923), 69–93.
9. For a lucid interpretation of Father Brown as a moralist and theologian, see John Peterson's "Father Brown's War on the Permanent Things," in *Permanent Things,* ed. A. A. Tadie and M. H. MacDonald (Grand Rapids, MI: William B. Eeerdmans, 1995).
10. Chesterton, "Blast of the Book," 81.
11. Ibid., 90–91.
12. Ibid., 91–92.

13. Charles. T. Onions, ed., *Oxford Dictionary of English Etymology* (London: Oxford University Press, 1966).
14. This name, along with the others used throughout this book, is a pseudonym.
15. This description and others cited in this work were collected from students and friends over a twenty-year period. For a discussion of the methodological issues involved in collecting and interpreting descriptions and of the basic nature of a phenomenological approach to psychology, see Chapter 5.
16. Gitta Sereny, *Albert Speer: His Battle with Truth* (New York: Alfred A. Knopf, 1995), 427–428.
17. Joseph E. Stiglitz, *Globalization and Its Discontents* (New York: Norton, 2002), 24.

Chapter 1

1. Clive S. Lewis, *Till We Have Faces* (Grand Rapids, MI: Eerdmans, 1966), 106.
2. Jeffrey Smith, *Where the Roots Reach the Water: A Personal and Natural History of Melancholia* (New York: North Point Press, 1999), 168.
3. Phil Mollon, Shame and Jealousy: The Hidden Turmoil (London: Karmac, 2002), p. 20.
4. L. A. Baxter and C. Bullis, "Turning Points in Romantic Relationships," *Communications Research* 12 (1986): 469–493. I am grateful to Professor Debra Sequeira, Department of Communications, Seattle Pacific University, for bringing this research tradition to my attention.
5. As I indicated in the Introduction, issues of method and the assumptions upon which this book is based are discussed in Chapter 5.
6. Parts of this chapter are based on Steen Halling, "Seeing a Significant Other 'As if for the First Time,'" in *Duquesne Studies in Phenomenological Psychology*, Vol. 3, ed. Amedeo Giorgi, Anthony Barton, and Charles Maes (Pittsburgh, PA: Duquesne University Press, 1983), 122–136.
7. One person provided a description of being disillusioned by a significant other. The relationship between "seeing the other as if for the first time" and being disillusioned will be discussed at the end of Chapter 2.
8. For a further discussion of the meaning of context, see Constance T. Fischer, "Personality and Assessment," in *Existential-Phenomenological Perspectives in Psychology*, ed. Ronald S. Valle and Steen Halling (New York: Plenum, 1989), esp. 161–163.
9. H. S. Sullivan, *The Interpersonal Theory of Psychiatry* (New York: Norton, 1953), 245.
10. Ibid., 167.
11. Louis S. Sass, "Humanism, Hermeneutics, and the Human Subject," in *Hermeneutics and Psychological Theory*, ed. S. B. Messer, L. A. Sass, and R. L. Woolfolk (New Brunswick, NJ: Rutgers University Press, 1988). Sass is summarizing (as well as agreeing with) the perspectives of modern hermeneutical

thinkers, in this section—specifically, the anthropologist Clifford Geertz and the philosopher Hans-Georg Gadamer.

12. Adrian van Kaam, *Existential Foundations of Psychology* (Pittsburgh, PA: Duquesne University Press, 1966), Chapter 10.

13. Ibid., 324–327.

14. Ibid., 326.

15. Martin Buber, *I and Thou,* trans. Walter Kauffman (New York: Scribner and Sons, 1970).

16. cf. Maurice Friedman's discussion of the I-Thou relationship in his book *Martin Buber: The Life of Dialogue* (New York: Harper Torchbooks, 1960).

17. Maurice Merleau-Ponty, *The Visible and the Invisible,* trans. Alfonso Lingis (Evanston, IL: Northwestern University Press, 1968).

18. Ibid., 10–11.

19. Max Scheler, *The Nature of Sympathy,* trans. Peter Heath (Howdon, CT: Archon Books, 1970).

20. Maurice Natanson, "Anonymity and Recognition. Toward an Ontology of Social Roles," in *Condito Humana,* cd. W. Van Bayer and R. M. Griffith (Berlin: Springer Verlag, 1966), 263.

21. Amedeo P. Giorgi, *Psychology as a Human Science* (New York: Harper and Row, 1970).

22. Emmanuel Levinas, *Totality and Infinity,* trans. Alfonso Lingis (Pittsburgh, PA: Duquesne University Press, 1969).

23. Maurice Merleau-Ponty, The *Phenomenology of Perception,* trans. Colin Smith (New York: Humanities Press, 1962).

24. One writer in the existential tradition who argues that we are fundamentally alone is the psychiatrist Irwin Yalom. See, for example, his *Existential Psychotherapy* (New York: Basic Books, 1980). In contrast, the French existential philosopher Gabriel Marcel takes the position that "the we is prior to the I."

25. Martin Buber, "Distance and Relation," in *The Knowledge of Man: The Philosophy of the Interhuman,* ed. Maurice Friedman (New York: Harper Torchbooks, 1965), 59–72.

26. Ibid., 69.

27. Gabriel Marcel, in particular, has challenged the equation of receptivity with passivity and emphasized the creative and responsive dimensions of receptivity. See, for example, his essay "Testimony and Existentialism," in *The Philosophy of Existentialism,* trans. Manya Harari (New York: Citadel Press, 1991), 91–103.

28. Martin Buber, "Elements of the Interhuman," *Psychiatry* 20, no. 2 (1957): 105–113. For a discussion of the place of imagination in relationships, see Steen Halling, "The Imaginative Constituent in Interpersonal Living: Empathy, Illusion and Will," in *Imagination and Phenomenological Psychology,* ed. Edward L. Murray (Pittsburgh, PA: Duquesne University Press, 1987), 140–174.

29. This is how William Lynch describes creativity. See his *Images of Hope: Imagination as Healer of the Hopeless* (Notre Dame, IN: University of Notre Dame Press, 1974).

30. Gabriel Marcel, "Towards a Phenomenology and a Metaphysics of Hope," in *Homo Viator: Introduction to a Metaphysics of Hope,* trans. Emma Craufurd (New York: Harper and Row, 1962), 29–67.

31. Walter T. Davis, *Shattered Dream: America's Search for Its Soul* (Valley Forge, PA: Trinity Press International, 1994), 165.

32. Marcel, "Toward a Phenomenology," 35.

33. Arthur Egendorf, *Healing from the War: Trauma and Transformation after Vietnam* (New York: Houghton Mifflin, 1985), 52.

34. Ibid.

35. Martin Heidegger, *Being and Time,* trans. John Macquarrie and Edward Robinson (New York: Harper and Row, 1962).

36. J. H. van den Berg, *A Different Existence* (Pittsburgh, PA: Duquesne University Press, 1972), 94.

Chapter 2

1. Erwin Straus, "Shame as a Histiological Problem," in *Phenomenological Psychology* (New York: Basic Books, 1966), 222.

2. Lin Bauer, Jack Duffy, Elizabeth Fountain, Steen Halling, Marie Holzer, Elaine Jones, Michael Leifer, and Jan O. Rowe, "Exploring Self-Forgiveness," *Journal of Religion and Health* 31, no. 2 (1992): 149–160; Jan O. Rowe, Steen Halling, Emily Davies, Michael Leifer, Diane, Powers, Jeanne van Bronkhorst, "The Psychology of Forgiving Another: A Dialogal Approach," in *Existential-Phenomenological Perspectives in Psychology,* ed. Ronald S. Valle and S. Halling (New York: Plenum, 1989), 179–192.

3. For a more detailed discussion of the method used in this chapter, see Chapter 5.

4. Bernd Jager, "Of Masks and Marks, Therapists, and Masters," *Journal of Phenomenological Psychology* 21, no. 2 (1992): 165–179.

5. Rollo May, *Power and Innocence: A Search for the Sources of Violence* (New York: Norton, 1972).

6. Ibid., 41.

7. Charles W. Socarides, "On Disillusionment: The Desire to Remain Disappointed," in *The World of Emotions,* ed. Charles. W. Socarides (New York: International Universities Press, 1977), 553–574.

8. Harold Searles, "The Psychodynamics of Vengefulness," *Psychiatry* 19, no. 1 (1956): 31–39.

9. Anna M. Antonovsky, "Idealization and the Holding of Ideals," *Contemporary Psychoanalysis* 27, no. 3 (1991): 389–404.

10. R. Janoff-Bulman, *Shattered Assumptions: Towards a New Psychology of Trauma* (New York: Free Press, 1992).

11. Melanie Klein, "On Observing the Behaviour of Young Infants" (1952), in *Melanie Klein: Envy and Gratitude and Other Works, 1946–1963* (New York: Delacorte, 1975), 97.

12. Constance F. Fischer, "Personality and Assessment," in *Existential-Phenomenological Perspectives in Psychology*, ed. Ronald S. Valle and Steen Halling (New York: Plenum, 1989), 157–178; Constance F. Fischer, *Individualizing Psychological Assessment* (Hillsdale, NJ: Lawrence Earlbaum, 1994).

13. For a lucid and careful overview of psychoanalytic theory, see Stephen A. Mitchell and Margaret J. Black, *Freud and Beyond: A History of Modern Psychoanalytic Thought* (New York: Basic Books, 1995).

14. See, for example, Meira Likierman, *Melanie Klein: Her Work in Context* (New York: Continuum, 2001).

15. Melanie Klein, "The Psycho-Analytic Play Technique," in *The Selected Melanie Klein*, ed. Juliet Mitchell (New York: Free Press, 1987), 35–54.

16. Chris Hedges, *War is a Force That Gives Us Meaning* (New York: Public Affairs, 2002).

17. Likierman, *Melanie Klein*, 192.

18. Ibid.

19. Klein, "Notes on Some Schizoid Mechanisms," in *Melanie Klein: Envy and Gratitude and Other Works*, 1–24.

20. Ibid.

21. Mitchell and Black, *Freud and Beyond*, 93.

22. Klein, "Notes on Some Schizoid Mechanisms."

23. Klein, "On Observing the Behavior of Young Infants," 112.

24. Mitchell and Black, *Freud and Beyond*.

25. Ibid.

26. Edith Jacobson, *The Self and the Object World* (New York: International Universities Press, 1964).

27. Ibid., 110–111.

28. Bas Levering, "The Language of Disappointment: On the Language Analysis of Feeling Words," *Phenomenology + Pedagogy* 10 (1992): 53–74, 71.

29. Jacobson, *The Self and the Object World*, 96.

30. Socarides, "On Disillusionment."

31. Ibid., 564.

32. Heinz Kohut, *The Kohut Seminars on Self Psychology and Psychotherapy with Adolescents and Young Adults*, ed. Miriam Elson (New York: Norton, 1987), 95.

33. Judith G. Teicholz, *Kohut, Loewald, and the Postmoderns: A Comparative Study of Self and Relationship* (Hillsdale, NJ: Analytic Press, 1999), Chapter 5.

34. Ibid.

35. See, for example, Heinz Kohut, *The Kohut Seminars*, in *The Search for the Self: Selected Writings of Heinz Kohut: 1950–1978*, Vols. 1 and 2., ed. Paul H. Ornstein (New York: International Universities Press, 1978).

36. Peter Homans, *The Ability to Mourn: Disillusionment and the Social Origin of Psychoanalysis* (Chicago: University of Chicago Press, 1989).

37. Ronnie Janoff-Bulman, *Shattered Assumptions: Towards a New Psychology of Trauma* (New York: Free Press, 1992), 70.

38. Ronnie Janoff-Bulman and Michael Berg, "Disillusionment and the Creation of Value: From Traumatic Losses to Existential Gains," *Perspectives on Loss: A Sourcebook,* ed. John H. Harvey (New York: Brunner/Mazel, 1998), 36.

39. Ibid., 44.

40. Vernon Holtz, "Being Disillusioned as Exemplified by Adults in Religion, Marriage, or Career: An Empirical Phenomenological Investigation" (PhD diss., Duquesne University, 1984); John J. Neubert, "Becoming and Being Disillusioned in Midlife: An Empirical Phenomenological Investigation" (PhD diss., Duquesne University, 1985); Christen Carson Daniels, "The Psychological Experience of Disillusionment in Young Adults: An Empirical Phenomenological Analysis" (PhD diss., Pacifica Graduate Institute, 2001).

41. Holtz, "Being Disillusioned," ii.

42. Neubert, "Becoming and Being Disillusioned," 129.

43. Silvano Arieti and Jules Bemporad, *Severe and Mild Depression* (New York: Basic Books, 1978).

44. Bauer et al., "Exploring Self-Forgiveness."

45. Thomas S. Eliot, *Four Quartets* (New York: Harcourt, Brace and Company, 1943), 3.

46. Paulus Berenson, *Finding One's Way with Clay: Pinched Pottery and the Color of Clay* (New York: Simon & Schuster, 1972), 21.

47. Maurice Merleau-Ponty, "The Child's Relation with Others," in *The Primacy of Perception and other Essays,* ed. James. M. Edie (Evanston, IL: Northwestern University Press, 1964), 108.

48. Gregory Baum, *Man Becoming: God in Secular Experience* (New York: Herder and Herder, 1971), 235.

Chapter 3

1. John Patton, *Is Human Forgiveness Possible?* (Nashville, TN: Abingdon Press, 1985), 16.

2. Laura Blumenfeld, *Revenge: A Story of Hope* (New York: Simon & Schuster, 2002), 126.

3. Steen Halling, "On Growing up as a Premodernist," in *Narrative Identities: Psychologists Engaged in Self-Construction,* ed. George Yancy and Susan Hadley (Philadelphia: Jessica Kingsley, 2005), 221.

4. Steen Halling, "Eugene O'Neill's Understanding of Forgiveness," in *Duquesne Studies in Phenomenological Psychology,* Vol. 3, ed. Amedeo Giorgi, Richard Knowles, and David L. Smith (Pittsburgh, PA: Duquesne University Press), 194.

5. Amedeo Giorgi, *Psychology as a Human Science* (New York: Harper and Row, 1970).

6. Hannah Arendt, *The Human Condition* (Chicago: University of Chicago Press, 1958).

7. Jan O. Rowe, Steen Halling, Michael Leifer, Emily Davies, Diane Powers, and Jeanne van Bronkhorst, "The Psychology of Forgiving Another: A Dialogal Research Approach," *Existential-Phenomenological Perspectives in Psychology*, ed. Ronald S. Valle and Steen Halling (New York: Plenum, 1989); Linn Bauer, Jack Duffy, Liz Fountain, Steen Halling, Marie Holzer, Elaine Jones, Michael, and Jan O. Rowe, "Exploring Self-Forgiveness," *Journal of Religion and Health* 31, no. 2: 149–160.

8. Valerie Fortney, "Hate or Heal," *Chatelaine,* August 1997: 54–57.

9. Ibid., 7.

10. Based on Rowe et al., "Psychology of Forgiving Another."

11. Part of what follows is taken from my article, "Embracing Human Fallibility: On Forgiving Oneself and Forgiving Others," *Journal of Religion and Health* 33, no. 2 (1994): 107–114. I want to thank David Leeming, the journal's editor, for permission to include it here.

12. Ibid., 109.

13. Ibid., 109.

14. J. Preston Cole, *The Problematic Self in Kierkegaard and Freud* (New Haven, CT; Yale University Press, 1971), 89.

15. Milo C. Milburn, "Forgiving Another: An Existential-Phenomenological Investigation" (PhD diss., Duquesne University, 1992), 177.

16. This dream is also mentioned in Bauer et al., "Exploring Self-Forgiveness."

17. Halling, "Embracing Fallibility," 110.

18. Ibid., 110.

19. Gabriel Marcel, "Sketch of a Phenomenology and a Metaphysics of Hope," in *Homo Viator,* trans. Emma Craufurd (New York: Harper and Row, 1962).

20. Leslie Farber, *Ways of the Will* (New York: Basic Books, 1966).

21. Ibid., 7.

22. Ibid., 15.

23. H. J. N. Horsburgh, "Forgiveness," *Canadian Journal of Philosophy* 4, no. 4 (1974): 271.

24. Milburn, *Forgiving Another.*

25. John Douglas Marshall, *Reconciliation Road: A Family Odyssey of War and Honor* (Seattle: University of Washington Press, 2000).

26. Ibid., 281.

27. Maurice Merleau-Ponty, *Phenomenology of Perception,* trans. Colin Smith (London: Routledge Kegan Paul, 1962), 420.

28. The earliest publication detailing the work and findings of this group was published in 1991: Robert D. Enright and the Human Development Group, "The Moral Development of Forgiveness," in *Handbook of Moral Behavior and Development,* Vol. 1, ed. W. Kurtines and J. Gewirtz (Hillsdale, NJ: Lawrence Earlbaum, 1991). See also Robert D. Enright, Elizabeth A. Gassin, and Ching Ru-Wu, "Forgiveness: A Developmental View," *Journal of Moral Education* 21, no. 2 (1992): 99–114.

29. Joanna North, "Wrongdoing and Forgiveness," *Philosophy* 42 (1987): 506.

30. Robert D. Enright and Richard P. Fitzgibbons, *Helping Clients Forgive: An Empirical Guide for Resolving Anger and Restoring Hope* (Washington, D.C.: American Psychological Association, 2000), 24.
31. Joanna North, "The Ideal of Forgiveness: A Philosopher's Exploration," in *Exploring Forgiveness,* ed. Robert D. Enright and Joanna North (Madison: University of Wisconsin Press, 1998), 15–34.
32. Robert Enright, Suzanne Freedman, and Julio Rique, "The Psychology of Interpersonal Forgiveness," in *Exploring Forgiveness,* ed. Robert D. Enright and Joanna North (Madison: University of Wisconsin Press), 50.
33. Suzanne R. Freedman and Robert Enright, "Forgiveness as Intervention Goal with Incest Survivors," *Journal of Consulting and Clinical Psychology* 64, no. 5 (1996): 983–992.
34. Martin E. Seligman, Elaine Walker, and David L. Rosenhan, *Abnormal Psychology,* 4th ed. (New York: Norton, 2001).
35. M. E. McCullough and E. L. Worthington, Jr., "Models of Interpersonal Forgiveness and their Application to Counseling: Review and Critique," *Counseling and Values* 39, no. 1 (1994): 4.
36. Enright and Fitzgibbons, *Helping Clients Forgive,* 16.
37. Elio Frattaroli, *Healing the Soul in the Age of the Brain: Becoming Conscious in an Unconscious World* (New York: Viking, 2001), 163.
38. William Barrett, *The Illusion of Technique: A Search for Meaning in a Technological Civilization* (Garden City, NY: Anchor Books, 1979).
39. Brian Keenan, *An Evil Cradling* (New York: Viking, 1993).
40. Ibid., 223.
41. Enright, Freedman, and Rique, "Psychology of Interpersonal Forgiveness."
42. Rowe et al., "Psychology of Forgiving Another," 242.

Chapter 4

1. Martin Buber, *I and Thou,* 2nd ed, trans. Ronald Gregor Smith (New York: Scribner, 1958), 11.
2. Robert Whitaker, *Mad in America: Bad Science, Bad Medicine, and the Enduring Mistreat of the Mentally Ill* (Cambridge, MA: Perseus, 2003), xiv.
3. David L. Kahn, "Reducing Bias," in *Hermeneutical Phenomenological Research: A Practical Guide for Nurse Researchers,* ed. Marlene Z. Cohen, David L. Kahn, and Richard H. Steeves (Thousand Oaks, CA: Sage, 2000), 89–90.
4. Gerald C. Davison, John M. Neale, Ann N. Kring, *Abnormal Psychology,* 9th ed. (New York: Wiley, 2004), 635.
5. Studies of institutionalization include John K. Wing and G. W. Brown, *Institutionalization and Schizophrenia: A Comparative Study of Three Mental Hospitals: 1960–1968* (Oxford: Cambridge University Press, 1970), and Edmond Phillips, "The Iatrogenic Environment: A Transactional Framework for Social Research," *Hospital & Community Psychiatry* 18, no. 2 (1967): 369–375. See also Gerald N. Grob, "Deinstitutionalization: The Illusion of

Policy," *Journal of Policy History* 9, no. 1 (1997): 49–73. I am grateful to my colleague Erica Lilleleht for bringing Grob's article to my attention.

6. Grob, "Deinstitutionalization."

7. Whitaker, *Mad in America,* has argued that major tranquilizers are far less effective than the claims made by their manufacturers and much of the psychiatric establishment. However, it is clear that suddenly going off such medication is apt to produce a relapse.

8. U.S. Department of Health and Human Services, *Mental Health: A Report of the Surgeon General* (Rockville, MD: U.S. Department of Health and Human Services, Substance Abuse and Mental Health Services Administration, Center for Mental Health Services, National Institutes of Health, and National Institutes of Mental Health, 1999).

9. Lee N. Robins, J. E. Helzer, M. M. Weissman, H. Orvaschel, E. Gruenberg, J. D. Burke, and D. A. Regier, "Lifetime Prevalence of Psychiatric Disorders in Three Sites," *Archives of General Psychiatry* 41 (1984): 949–958. This trend toward increase in depression is not restricted to the United States but has also been found, with some variation, in locations as far apart as Italy, Lebanon, New Zealand, and Taiwan, as well as in other countries. See, Cross-National Collaborative Group, "The Changing Rate of Major Depression: Cross-National Comparisons," *Journal of the American Medical Association* 268, no. 2 (1992): 3098–3105.

10. U.S. Department of Health and Human Services, *Mental Health.*

11. George H. Darby, Elliott T. Barker, and Michael H. Mason, "Buber behind Bars," paper presented at the Fall Meeting of the Ontario Psychiatric Association, October, 1966. Buber's overall influence on the field of psychotherapy has been substantial. For an overview of this issue, see Maurice S. Friedman, *The Healing Dialogue in Psychotherapy* (New York: Aronson, 1985), and Richard Hycner, *Between Person and Person: Toward a Dialogical Psychotherapy* (Highland, NY: Gestalt Journal, 1993).

12. The section that follows is based, in part, on a chapter I wrote with Judy Dearborn Nill, "Demystifying Psychopathology: Understanding Disturbed Persons," in *Existential-Phenomenological Perspectives in Psychology,* ed. Ronald S. Valle and Steen Halling (New York: Plenum, 1989): 179–192.

13. James Waller, *Becoming Evil: How Ordinary People Commit Genocide and Mass Killings* (New York: Oxford University Press, 2002), 202.

14. Mary Warnock, *Memory* (Faber and Faber: London, 1987), 130.

15. Julie Sharif, "Impatience as a Hindrance in Psychotherapy," unpublished paper, Seattle University (1984), 19.

16. Sigmund Freud, *The Psychopathology of Everyday Life,* trans. Alan Tyson (New York: Norton, 1965).

17. Barry N. Kaufman, *A Miracle to Believe In* (New York: Doubleday, 1980).

18. Oliver Sacks, *The Man who Mistook His Wife for a Hat and other Clinical Tales* (New York: Harper and Row, 1987): 92–101.

19. Ibid., 98.

20. Donald J. Cohen, "Into Life: Autism, Tourette's Syndrome and the Community of Clinical Research," in *Life Is with Others: Selected Writings on Child Psychiatry,* ed. Donald J. Cohen, Martin Andrés, and Roberta A. King (New Haven: Yale University Press, 2006), 15–25.

21. George E. Atwood and Robert D. Stolorow, "Experience and Conduct," *Contemporary Psychoanalysis* 17, no. 2 (1981): 197–208.

22. Faith A. Robinson, "Dissociative Women's Experiences of Self-Cutting," in *Phenomenological Inquiry in Psychology: Existential and Transpersonal Dimensions,* ed. Ronald S. Valle (New York: Plenum, 1998), 209–226.

23. Silvano Arieti, *Interpretation of Schizophrenia* (New York: Basic Books, 1974), 147–152.

24. Ibid., 148.

25. Courtney M. Harding, Joseph Zubin, and John S. Strauss, "Chronicity in Schizophrenia: Revisited," *British Journal of Psychiatry* 161 (1992): 27–37.

26. Melitta J. Leff, John F. Roatch, and William E. Bunney, Jr., "Environmental Factors Preceding the Onset of Severe Depression," *Psychiatry* 33 (1970): 293–311.

27. Harry S. Sullivan, *The Psychiatric Interview* (New York: Norton, 1954).

28. See, for example, Frieda Fromm-Reichman, *Psychoanalysis and Psychotherapy: Selected Papers* (Chicago: University of Chicago Press, 1959); Betram P. Karon and Gary VandenBos, *Psychotherapy of Schizophrenia: The Treatment of Choice* (Northvale, NJ: Arsonson, 1981); R. D. Laing, *The Divided Self* (Baltimore: Pelican, 1965); Garry Prouty, *Theoretical Evolutions in Person-Centered/Experiential Therapy: Applications to Schizophrenia and Retarded Psychoses* (Westwood, CT: Praeger, 1994).

29. Fromm-Reichman, *Psychoanalysis and Psychotherapy,* 126.

30. "A Recovering Patient, 'Can We Talk?' The Schizophrenic Patient in Psychotherapy," *American Journal of Psychotherapy* 143, no. 1 (1986): 70.

31. See, for example, Davison, et al., *Abnormal Psychology,* 284–286; Martin E. P. Seligman, Elaine F. Walker, and David Rosenhan, *Abnormal Psychology,* 4th ed. (New York: Norton, 2001), 265–267.

32. U.S. Department of Health and Human Services, *Mental Health;* Nancy Andreasen, "Linking Mind and Brain in the Study of Mental Illnesses: A Project for a Scientific Psychopathology," *Science* 275, March 14, 1997: 1586.

33. Elio Frattaroli, *Healing the Soul in the Age of the Brain: Becoming Conscious in an Unconscious World* (New York: Viking, 2001), 164; Amedeo P. Giorgi, *Psychology as a Human Science* (New York: Harper and Row, 1970).

34. American Psychiatric Association, *Diagnostic and Statistical Manual of Mental Disorders,* 4th ed., Text Revision (Washington, D.C.: Author, 2000), 305, 309.

35. Nathaniel Lachenmeyer, *The Outsider: A Journey into My Father's Struggle with Madness* (New York: Broadway Books, 2000), 251.

36. See the letter to the editor of the *American Journal of Psychiatry* 153, no. 10 (1996): 1373–1374 by Laura Lee Hall and Laurie M. Flynn, both directors at

NAMI, in response to an article by Leon Eisenberg, entitled "The Social Construction of the Human Brain," in the same journal (152, no. 11 (1995): 1563–1575). Hall and Flynn accuse Eisenberg of starting his article with an insult to the families of people with serious mental illness, an accusation that in my mind is something of a stretch since they are responding to one sentence in Eisenberg's article with little attention to the context. Eisenberg, in his response to their letter (October 1996, p. 1374), bends over backward to apologize, but also takes NAMI to task for being exclusively focused on the biological aspects of schizophrenia and other mental disorders. This exchange supports Eisenberg's (1995) contention in his article that there is a "highly charged public discussion on whether mental diseases are biological or psychological" (p. 1563).

37. Abigail Zuger, "Drug Companies' Sales Pitch: 'Ask your Doctor,'" *New York Times,* August 5, 1997, sec. C.

38. Robert Pear, "Drug Companies Increase Spending on Efforts to Lobby Congress and Governments," *New York Times,* June 1, 2003, sec. A.

39. Sheryl Gay Stolberg and Gardiner Harris, "Measure to Ease Imports of Drugs is Gaining in House," *New York Times,* July 22, 2003, sec. A.

40. Cynthia Crossen, *Tainted: The Manipulation of Fact in America* (New York: Touchstone/Simon & Schuster, 1996), 183.

41. Robert Pear, "Spending on Prescription Drugs Increases almost 19 Percent." *New York Times* May 8, 2001, A1.

42. Sophia F. Dziegielewski, *Psychopharmacology Handbook* (New York: Norton 2006).

43. Shankar Vedantam, "More Kids Receiving Psychiatric Drugs; Question of 'Why' still Unanswered," *Washington Post,* January 14, 2003, sec. A.

44. David Karp, *Living with Sadness: Depression, Disconnection, and the Meanings of Illness* (New York: Oxford University Press, 1996), 101.

45. Gerald I. Klerman, "The Advantages of DSM-III," *American Journal of Psychiatry* 141, no. 4 (1984): 542.

46. American Psychiatric Association (2000), *Diagnostic and Statistical,* xxx–xxxi.

47. George E. Vaillant, "The Disadvantages of DSM-III Outweigh its Advantages," *American Journal of Psychiatry* 141, no. 4 (1984): 542–545; Theodore Millon, "Classification in Psychopathology; Rationale, Alternatives and Standards," *Journal of Abnormal Psychology* 100, no. 3 (1991): 245–261; John Mirowsky and Catherine E. Ross, "Psychiatric Diagnosis as Reified Measurement," *Journal of Health and Social Behavior* 30 (1989): 11–25.

48. Ellen Corin and Gilles Lauzon, "Positive Withdrawal and the Quest for Meaning: Reconstruction of Experience among Schizophrenics," *Psychiatry* 55, no. 1 (1992), 267.

49. Quoted in S. A. Kirk and H. Hutchins, *The Selling of DSM: The Rhetoric of Science in Psychiatry* (Hawthorne, NY: Aldine de Gruyter, 1992), 248.

50. Philip Cushman, "How Psychology Erodes Personhood," *Journal of Theoretical and Philosophical Psychology* 22, no. 2 (2002): 108.

51. Millon, "Classification in Psychopathology," 255.
52. Ibid., 11.
53. Whitaker, *Mad in America.*
54. David H. Barlow, "The Effectiveness of Psychotherapy: Science and Policy," *Clinical Psychology: Science and Practice* 3, no. 3 (1996): 236–240.
55. Diane Chambless, "In Defense of Dissemination of Empirically Supported Psychological Interventions," *Clinical Psychology: Science and Practice* 3, no. 3 (1996): 232.
56. Sol G. Garfield, "Some Problems associated with 'Validated' Forms of Psychotherapy," *Clinical Psychology: Science and Practice* 3, no. 3 (1996): 218–229; Arthur. C. Bohart, Maureen O'Hara, and Larry. M. Leitner, "Empirically Violated Treatments: Disenfranchisement of Humanistic and other Psychotherapies," *Psychotherapy Research* 8, no. 4 (1998): 141–157.
57. William Barrett, *The Illusion of Technique: A Search for Meaning in a Technological Civilization* (Garden City, NY: Anchor Books, 1979).
58. Ursala Franklin, *The Real World of Technology* (Concord, Canada: Anansi, 1992).
59. David F. Noble, *The Religion of Technology: The Divinity of Man and the Spirit of Invention* (New York: Alfred A. Knopf, 1998).
60. Erica Lilleleht, "The Paradox of Practice: Using Rhetoric to Understand the Dilemmas of Psychiatric Rehabilitation," *International Journal of Psychosocial Rehabilitation* 10, no. 1 (2005). http://www.psychosocial.com/IJPR_10/Paradox_in_Practice_Lilleleht.html.
61. Ibid.
62. U.S. Department of Health and Human Services, *Mental Health,* 454.
63. Constance T. Fischer, "Personality and Assessment," in *Existential-Phenomenological Perspectives in Psychology,* ed. Ronald S. Valle and Steen Halling (New York: Plenum), 168. See also Constance T. Fischer, *Indvidualizing Psychological Assessment* (Hillsdale, NJ: Lawrence Earlbaum, 1995).
64. Frattaroli, *Healing the Soul.* 62.
65. Andreasen, "Linking Mind and Brain," 1587.
66. Elliot S. Valenstein, *Blaming the Brain: The Truth about Drugs and Mental Health* (New York: Free Press, 1998), 125.
67. William Bevan, "Contemporary Psychology: A Tour inside the Onion," *American Psychologist* 46, no. 5 (1991): 476.
68. Joanne Greenberg, *I Never Promised You a Rose Garden* (New York: New American Library, 1964); Sylvia Nasar, *A Beautiful Mind: The Life of Mathematical Genius and Nobel Laureate John Nash* (New York: Simon & Schuster, 1998).

Chapter 5

1. David A. Karp, *Speaking of Sadness: Depression, Disconnection and the Meaning of Illness.* (New York: Oxford University Press, 1996), 202.

2. William Bevan, "A Tour inside the Onion," *American Psychologist* 46, no. 5 (1991): 478.

3. Edward E. Jones and Harold B. Gerard, *Foundations of Social Psychology* (New York: Wiley & Sons, 1967), 33–35.

4. Paul Lazarsfeld's review is entitled, "The American Soldier—An Expository Review," *Public Opinion Quarterly* 13, no. 2 (1949): 377–404.

5. Michael W. Passer and Ronald E. Smith discussed the same issue in their *Psychology: Frontiers and Applications* (New York: McGraw-Hill, 2001), 45–47.

6. Robert D. Romanyshyn and Brian J. Whalen, "Psychology and the Attitude of Science," in *Existential-Phenomenological Perspectives in Psychology,* ed. Ronald S. Valle and Steen Halling (New York: Plenum, 1989), 17–39.

7. Maurice Merleau-Ponty, *Phenomenology of Perception,* trans. Colin. Smith (New York: Humanities Press, 1962), viii.

8. Miriam Waddington, in *Call Them Canadians,* ed. Lorraine Monk (Ottawa, Canada: Queen's Printer, 1968), 229. Included with the kind permission of Jonathan Waddington.

9. Inger Alver Gløersen, *Den Munch Jeg Møtte,* 2nd ed. [The Munch I met] (Oslo: Gyldendal Norsk, 1962).

10. Helen Valentine, ed., *Art in the Age of Queen Victoria: Treasures from the Royal Academy of Arts Permanent Collection* (London: Royal Academy of Arts, 1999). The painting *On Strike* is shown on p. 121 of Valentine's book.

11. George Steiner, *Real Presences* (Chicago: University of Chicago Press, 1991), 226.

12. Kathleen Norris, *Amazing Grace: A Vocabulary of Faith* (New York: Riverhead Books, 1998), 17.

13. For an overview of the influence of phenomenological philosophy on psychiatry and psychology, see Steen Halling and Judy Dearborn Nill, "A Brief History of Existential-Phenomenological Psychiatry and Psychotherapy," *Journal of Phenomenological Psychology* 26, no. 1 (1995): 1–45.

14. Walter Lowrie, *A Short Life of Kierkegaard* (Princeton, NJ: Princeton University Press, 1970), 115.

15. Søren Kierkegaard, *The Sickness unto Death,* trans. and ed. Walter. Lowrie (Princeton, NJ: Princeton University Press, 1941), 29.

16. Gabriel Marcel, "An Essay in Autobiography," in *The Philosophy of Existentialism,* trans. Manya Harari (New York: Citadel Press, 1991), 121.

17. Gabriel Marcel, "Towards a Phenomenology and a Metaphysics of Hope," in *Homo Viator: Introduction to a Metaphysics of Hope,* trans. Emma Craufrud (New York: Harper and Row, 1974), 29–67.

18. Peter T. Manicas and Paul F. Secord, "Implications of the New Philosophy of Science," *American Psychologist* 38 (1983): 399–413.

19. Thomas S. Kuhn, *The Structure of Scientific Revolution* (Chicago: University of Chicago Press, 1962), rev. ed., 1970.

20. Michael Polanyi, *Personal Knowledge: Toward a Post-Critical Philosophy* (Chicago: University of Chicago Press, 1958).

21. Thomas Kuhn discusses this issue at length in "Second Thoughts on Paradigms," Chapter 12, in his book, *The Essential Tension: Selected Studies in Scientific Tradition and Change* (Chicago: University of Chicago Press, 1977).

22. Stephen Jay Gould, *The Hedgehog, the Fox, and the Magister's Pox: Mending the Gap between Science the Humanities* (New York: Harmony Books, 2003), 114.

23. Stephen Toulmin and David E. Leary, "The Cult of Empiricism in Psychology and Beyond," in *A Century of Psychology as a Science*, ed. Sigmund Koch and David E. Leary (New York: McGraw-Hill, 1985), 609.

24. Kuhn, *Structure of Scientific Revolution* (1970), 162–163.

25. Susan Nolen-Hoeksema, *Abnormal Psychology*, 3rd ed. (New York: McGraw Hill, 2004), 69.

26. Manicas and Secord, "Implications of the New Philosophy," 410.

27. Nolen-Hoeksema, *Abnormal Psychology*, xiii.

28. Edmund Husserl, *The Crisis of European Sciences and Transcendental Phenomenology*, trans. David Carr (New York: Humanities Press, 1970). See also Herbert Spiegelberg, *The Phenomenological Movement*, 3rd ed. (The Hague: Martinus Nijhoff, 1982).

29. Spiegelberg, *The Phenomenological Movement*, 747.

30. Toulmin and Leary, "Cult of Empiricism."

31. Halling and Dearborn Nill, "A Brief History"; Amedeo P. Giorgi, *Psychology as a Human Science: A Phenomenologically Based Approach* (New York: Harper and Row, 1970); William James, *Varieties of Religious Experience* (New York: Modern Library, 1963); Eugene I. Taylor, Jr., "On Psychology as a Person-Centered Science: Williams James and His Relation to the Humanistic Tradition," in Donald Moss, ed., *Humanistic and Transpersonal Psychology: A Historical and Biographical Sourcebook* (Westport, CT: Greenwood Press, 1999), 301–313.

32. Giorgi, *Psychology as a Human Science*.

33. Frederick J. Wertz and Christopher Aanstoos, "Amedeo Giorgi and the Project of a Human Science," in *Humanistic and Transpersonal Psychology*, ed. Donald Moss, (Westport, CT; Greenwood Press), 287–300; David L. Smith, *Fearfully and Wonderfully Made: The History of Duquesne University's Graduate Program (1959–1999)* (Pittsburgh: Simon Silverman Phenomenology Center, Duquesne University, 2002).

34. Giorgi, *Psychology as a Human Science*, 126.

35. Charles T. Onions, ed., *The Oxford Dictionary of English Etymology* (London: Oxford University Press, 1966), 797.

36. Sylvia Nasar, *A Beautiful Mind: The Life of Mathematical Genius and Nobel Laureate John Nash* (New York: Touchstone/Simon & Schuster, 2001), 368.

37. See, for example, Gerald C. Davison, John M. Neale, and Ann M. Kring, *Abnormal Psychology*, 9th ed. (New York: Wiley, 2004).

38. *Random House Webster's College Dictionary* (New York: Random House, 2001), 1178.

39. See Toulmin and Leary, "Cult of Empiricism," and Manicas and Secord, "Implications of the New Philosophy," for a discussion of the issue of "natural law" and psychology.

40. See, for example, Karin Dahlberg, Nancy Drew, and Maria Nyström, *Reflective Lifeworld Research* (Lund, Sweden: Studdentlitteratur, 2001); Constance T. Fischer, ed., *Qualitative Research Methods for Psychologists: Introduction through Empirical Studies* (New York: Academic Press, 2006); Amedeo Giorgi, ed., *Phenomenology and Psychological Research* (Pittsburgh: Duquesne University Press, 1985); Gunnar Karlsson, *Psychological Qualitative Research from a Phenomenological Perspective* (Stockholm: Almqvist & Wiksell, 1993); Clark Moustakas, *Phenomenological Research Methods* (Thousand Oaks, CA: Sage, 1994); Clark Moustakas, *Heuristic Research: Design, Methodology and Applications* (Thousand Oaks, CA: Sage,1990); Steinar Kvale, *InterViews: An Introduction to Qualitative Research Interviewing* (Thousand Oaks, CA: Sage, 1996); Jonathan Smith, ed., *Qualitative Psychology: A Practical Guide to Research Methods* (Thousand Oaks, CA: 2003). See also the "Special Edition on Methodology" of the *Indo-Pacific Journal of Phenomenology* 6 (2006). http://www.ipjp.org/SEmethod/index.html.

41. Spiegelberg, *Phenomenological Movement*.

42. Brittney Beck, Steen Halling, Marie McNabb, Daniel Miller, Jan O. Rowe, and Jennifer Schulz, "Facing up to Hopelessness: A Dialogal Phenomenological Approach," *Journal of Religion of Health* 42, no. 4 (2003): 339–354.

43. The dialogal method is described in some detail in the following articles: Steen Halling and Michael Leifer, "The Theory and Practice of Dialogal Research," *Journal of Phenomenological Psychology* 22, no.1 (1991): 1–15; Steen Halling, George Kunz, and Jan O. Rowe, "The Contributions of Dialogal Psychology to Phenomenological Research," *Journal of Humanistic Psychology* 34, no.1 (1994): 109–131; Steen Halling, Michael Leifer, and Jan O. Rowe, "Emergence of the Dialogal Approach: Forgiving Another," in *Qualitative Research Methods for Psychologist*, ed. Constance T. Fischer (New York: Academic Press, 2006), 247–278.

44. Amedeo Giorgi, "An Application of Phenomenological Method in Psychology," in *Duquesne Studies in Phenomenological Psychology*, Vol. 2, ed. A. Giorgi, C. Fischer, and E. Murray (Pittsburgh: Duquesne University Press, 1975), 84.

45. Constance T. Fischer and Frederick J. Wertz, "Empirical Phenomenlogical Analyses of Being Criminally Victimized," in *Duquesne Studies in Phenomenological Psychology*, Vol. 3, ed. A. Giorgi, R. Knowles and D. L. Smith (Pittsburgh: Duquesne University Press, 1979), 135–158.

46. Ibid., 144.

47. Giorgi, "Application of Phenomenological Method," 89.

48. Ibid., 89.

49. Ibid., 92.

50. Ibid., 94–95.

51. Hans-Georg Gadamer, *Truth and Method* (New York: Crossroads, 1975), 345.

52. Jan O. Rowe, Steen Halling, Emily Davies, Michael Leifer, Dianne Powers, and Jeanne van Bronkhorst, "The Psychology of Forgiving Another: A Dialogal Research Approach," in *Existential-Phenomenological Perspectives*, ed. Valle and Halling, 233–244.

53. Eugene T. Gendlin, "Experiential Phenomenology," in *Phenomenology and the Social Sciences*, ed. Maurice Natanson (Evanston, IL: Northwestern University Press, 1973), 294.

54. George Kunz, Daniel Clingaman, Renee Hulet, Richard Kortsep, Bruce Kugler, and Mineko-Sung Park, "A Dialogal Phenomenological Study of Humility" (paper presented at the Sixth International Human Science Research Conference, University of Ottawa, Canada, June 1987).

55. Michael Leifer, "The Dialogal Approach to Phenomenological Research" (paper presented at the Fifth International Human Science Research Conference, University of California at Berkeley, June 1986).

56. George Gusdorf, *Speaking*, trans. Paul. T. Brockelman (Evanston, IL: Northwestern University Press, 1965), 116.

57. Stephen Jay Gould, *The Hedgehog*, 34.

58. See Ronald S. Valle, Mark King, and Steen Halling, "An Introduction to Existential-Phenomenological Thought in Psychology," in *Existential-Phenomenological Perspectives in Psychology*, ed. Valle and Halling, especially 10–11, and Giorgi, *Psychology as a Human Science*, 144–155.

59. William F. Fischer, "An Empirical-Phenomenological Investigation of Being Anxious: An Example of the Phenomenological Approach to Emotion," in *Existential-Phenomenological Perspectives*, ed. Valle and Halling, 127–136.

60. Ibid, 134.

61. Spiegelberg, Phenomenological Movement; Giorgi, *Psychology as a Human Science*.

62. Clifford Geertz, quoted in Gould, *The Hedgehog*, 156.

63. Robert Elliott, Constance T. Fischer, and David L. Rennie, "Evolving Guidelines for Publication of Qualitative Research Studies in Psychology and Related Fields," *British Journal of Clinical Psychology* 38 (1999): 215–229.

64. Lisa L. S. Tsoi Hoshmand, "Alternative Research Paradigms: A Review and a Teaching Proposal," *The Counseling Psychologist* 17, no. 1 (1989): 3–79. See also, Donald E. Polkinghorne, *Methodology for the Human Sciences: Systems of Inquiry* (Albany: State University of New York Press, 1983), as well as his "Phenomenological Research Methods," in *Existential-Phenomenological Perspectives*, ed. Valle and Halling, 41–60.

65. Lane A. Gerber, *Married to their Careers: Career and Family Dilemmas* (New York: Tavistock, 1983).

66. Marc Briod, "The Young Child's Sense of Time and the Clock," *Phenomenology + Pedagogy* 4, no. 1 (1986): 9–19.

67. I. Lundgren and Karin Dahlberg, "Women's Experience of Pain during Childbirth," *Midwifery* 14, no. 2 (1998): 105–110.

68. Sandra M. Levy, "The Experience of Undergoing a Heart Attack: The Construction of a New Reality," *Journal of Phenomenological Psychology* 12, no. 2 (1981): 153–171.
69. Larry Davidson, L. Hoge, M. A. Merrill, J. Rakfeldt, and E. E. H. Griffith, "The Experiences of Long-Stay Inpatients Returning to the Community," *Psychiatry* 58, no. 2 (1995): 122–132.
70. Herman Coenen, "Improvised Contexts: Movement, Perception and Expression in Deaf Children's Interaction," *Journal of Phenomenological Psychology* 17, no. 1 (1986): 1–31.
71. Harry J. Berman, *Interpreting the Aging Self* (New York: Springer, 1994).
72. There are several web sites that list such studies (see http://www.artfulsoftware.com/humansciencesearch.html and http://www.phenomenologyonline.com).

Chapter 6

1. Donnel B. Stern, "Unformulated Experience: From Familiar Chaos to Creative Disorder," *Contemporary Psychoanalysis* 12, no. 1 (1983): 92.
2. John O'Donohue, *Beauty: The Invisible Embrace* (New York: HarperCollins, 2004), 132.
3. Amedeo Giorgi, "The Search for the Psyche: A Human Science Perspective," in *The Handbook of Humanistic Psychology. Leading Edges in Theory, Research and Practice,* ed. Kirk J. Schneider, James F. Bugental, and J. Fraser. Pierson (Thousand Oaks, CA: Sage, 2001), 62.
4. B. F. Skinner, *Beyond Freedom and Dignity* (New York: Knopf, 1971).
5. Kenneth J. Gergen, "The Social Constructionist Movement in Modern Psychology," *American Psychologist* 20 (1985): 266–275; Steen Halling and Charles Lawrence, "Social Constructionism: Homogenizing the World, Negating Embodied Experience," *Journal of Theoretical and Philosophical Psychology* 19, no.1 (1999): 78–89.
6. Daniel J. Goldhagen, *Hitler's Willing Executioners: Ordinary Germans and the Holocaust* (New York: Knopf, 1996), 34.
7. Steen Halling, "On Growing up as a Premodernist,'" in *Narrative Identities: Psychologists Engaged in Self Construction,* ed. George Yancy and Susan Hadley (London: Jessica Kingsley, 2005), 221.
8. Martin Heidegger, *Being and Time,* trans. John Macquarrie and Edward Robinson (New York: Harper & Row, 1962).
9. Donald J. Cohen, "Into Life: Autism, Tourette's Syndrome and the Community of Clinical Research," in *Life is with Others: Selected Writings on Child Psychiatry,* ed. Donald Cohen, Martin Andrés, and Roberta A. King (New Haven, Yale University Press, 2006), 24. I want to thank my colleague Jeanette Valentine for bringing Cohen's lecture to my attention.
10. Ibid., 25.
11. Ibid., 26.

12. James E. Faulconer, ed., *Transcendence in Philosophy and Religion* (Bloomington: Indiana University Press, 2003).
13. James Faulconer, "Introduction: Thinking Transcendence," in Faulconer, ibid., 7.
14. Emmanuel Levinas, *Totality and Infinity*, ed. Alfonso Lingis (Pittsburgh: Duquesne University Press, 1969); Marlène Zarader, "Phenomenality and Transcendence," in *Transcendence in Philosophy and Religion*, ed. James E. Faulconer (Bloomington: Indiana University Press, 2003): 106–119.
15. Faulconer, ed., *Transcendence in Philosophy and Religion*.
16. Zarader, "Phenomenality and Transcendence," 109–110.
17. Ernest Becker, *The Denial of Death* (New York: Free Press, 1973), ix.
18. Ernest Becker, *Escape from Evil* (New York: Free Press, 1975), 4.
19. Ibid., 64.
20. Ibid., 5.
21. Huston Smith, *Beyond the Post-Modern Mind* (New York: Crossroads, 1982), 95.
22. Terry Eagleton, *After Theory* (New York: Basic Books, 2003), 212.
23. Ibid., 51.
24. Rollo May, *Power and Innocence: A Search for the Sources of Violence* (New York: Norton, 1972), 23.
25. Tom Pyszczynski, Sheldon Solomon, and Jeff Greenberg, *In the Wake of 9/11: The Psychology of Terror* (Washington, D.C.: American Psychological Association, 2003).
26. Sheldon Solomon, Jeff Greenberg, and Tom Pyszczynski, "Tales from the Crypt: On the Role of Death in Life," *Zygon* 33 (1998): 9–43.
27. James Waller, *Becoming Evil: How Ordinary People Commit Genocide and Mass Killings* (New York: Oxford University Press, 2002).
28. Daniel Liechty, *Transference and Transcendence: Ernest Becker's Contribution to Psychotherapy* (New York: Aronson, 1995), 169.
29. Glenn Hughes, *Transcendence and History: The Search for Ultimacy from Ancient Societies to Postmodernity* (Columbia: University of Missouri Press, 2003), 210.
30. Interview with Ernest Becker by Sam Keen. The quote is taken from a corrected transcript of the original interview that Neil Elgee, president of the Ernest Becker Foundation, was kind enough to give me. The quotes are from pages 18 and 17 of this transcript, respectively. An abbreviated version of the interview appeared in *Psychology Today*, April (1974), under the title, "The Heroics of Everyday Life: A Theorist of Death Confronts His Own End: A Conversation with Ernest Becker by Sam Keen," 71–80.
31. Lance Morrow, *Evil: An Investigation* (New York: Basic Books, 2003), 8.
32. Charles Pope, "Study: Parents, Government Need to Fight Child Obesity," *Seattle Post-Intelligencer*, October 1, 2004, sec. A.
33. Allen Wheelis, *How People Change* (New York: Harper Torchbooks, 1974), 88.
34. Maurice Merleau-Ponty, *Phenomenology of Perception*, trans. Colin Smith (New York: Humanities Press, 1962).

35. George Kunz, *The Paradox of Power and Weakness: Levinas and an Alternative Paradigm for Psychology* (Albany, NY: State University Press of New York, 1998).

36. Richard H. Cox, "Transcendence and Imminence in Psychotherapy," *American Journal of Psychotherapy* 51, no. 4 (1997): 511. Parts of this chapter are based on my article, "Meaning beyond Heroic Illusions? Transcendence in Everyday Life," *Journal of Religion and Health* 39, no. 2 (2000): 143–157.

37. Cox, "Transcendence and Imminence in Psychotherapy," 514.

38. Abraham H. Maslow, "Lessons from the Peak-Experiences," *Journal of Humanistic Psychology* 2, no. 1 (1962): 9.

39. Abraham H. Maslow, "Various Meanings of Transcendence," *Journal of Transpersonal Psychology* 1 (1969): 56–66.

40. Jenny Wade, *Transcendent Sex: When Lovemaking Opens the Veil* (New York: Simon & Schuster, 2004).

41. Merleau-Ponty, *Phenomenology of Perception*, 376.

42. Ibid., 382.

43. Martin Heidegger, *The Basic Problems of Phenomenology,* trans. Albert Hofstadter (Bloomington: Indiana University Press, 1982), 297.

44. Ibid., 298.

45. Ibid., 299.

46. W. V. Dych, "Theology in a New Key," in *A World of Grace: An Introduction to the Themes and Foundations of Karl Rahner's Theology,* ed. I. Journal O'Donovan (New York: Seabury Press, 1980), 1–16.

47. Karl Rahner, "The Experience of God Today," in *Theological Reflections,* Vol. 11, trans. David Bourke (New York: Seabury, 1974), 174.

48. James Jones, *Contemporary Psychoanalysis and Religion: Transference and Transcendence* (New Haven, CT: Yale University Press, 1991), 125.

Chapter 7

1. Maurice Merleau-Ponty, *The Visible and the Invisible,* trans. Alfonso Lingis (Evanston, IL: Northwestern University Press, 1968), 104.

2. Karen Armstrong, *The Spiral Staircase: My Climb out of Darkness* (New York: Knopf, 2004), 279.

3. George Frank, "The Intersubjective School of Psychoanalysis: Concerns and Questions," *Psychoanalytic Psychology* 15, no. 3 (1998): 420–423.

4. Karen J. Prager, *The Psychology of Intimacy* (New York: Guilford, 1995), 1.

5. Debra J. Mashek and Arthur Aron, eds., *Handbook of Closeness and Intimacy* (Mahwah, NJ: Lawrence Erlbaum, 2004), 1.

6. Mashek and Aron, *Handbook,* 2.

7. Prager, *The Psychology of Intimacy,* 17.

8. Robert A. Baron and Donne Byrne, *Social Psychology,* 8th ed. (Boston: Allyn and Bacon, 1997), 7.

9. Steve Duck, *Meaningful Relationships: Talking, Sense and Relating* (Thousand Oaks, CA: Sage, 1994), 34.

10. Fritz Heider, *The Psychology of Interpersonal Relations* (New York: Wiley, 1958).

11. Duck, *Meaningful Relationships*, 6.

12. Arthur P. Aron, Debra J. Mashek, and Elaine N. Aron, "Closeness as Including Others in Self," in *Handbook of Closeness and Intimacy*, ed. Debra J. Mashek and Arthur P. Aron (Mahwah, NJ: Lawrence Earlbaum, 2004), 28.

13. John W. Thibaut and H. H. Kelley, *The Social Psychology of Groups* (New York: Wiley, 1959).

14. Thus, Prager, *The Psychology of Intimacy*, 4, writes of how, in self-disclosure, we open ourselves up to being influenced by the other.

15. Jean-Philippe Laurenceau, Luis M. Rivera, Amy R. Schaffer, and Paula R. Pietromonaco, "Intimacy as an Interpersonal Process: Current Status and Future Directions," in *Handbook of Closeness and Intimacy*, ed. Debra J. Mashek and Arthur P. Aron (Mahwah, NJ: Lawrence Earlbaum, 2004), 61–80.

16. Lisa M. Register and Tracy B. Henley, "The Phenomenology of Intimacy," *Journal of Social and Personal Relationships* 9 (1992): 467–481.

17. Karen J. Prager and Linda J. Roberts, "Deep Intimate Connection: Self and Intimacy in Couple Relationships," in *Handbook of Closeness and Intimacy*, ed. Debra J. Mashek and Arthur P. Aron (Mahwah, NJ: Lawrence Earlbaum and Associates, 2004), 44.

18. Lewis Aron, "On the Unique Contribution of the Interpersonal Approach to Interaction, *Contemporary Psychoanalysis* 41, no. 1 (2005): 21–34.

19. Robert Stolorow and George Atwood, *Faces in a Cloud: Subjectivity in Personality* Theory (New York: Aronson, 1979), 11.

20. Robert D. Stolorow, George E. Atwood, and Donna M. Orange, *Worlds of Experience: Interweaving Philosophical and Clinical Dimensions in Psychoanalysis* (New York: Basic Books, 2002).

21. Stolorow and Atwood, *Faces in a Cloud*, 11.

22. Ibid., 183.

23. Harry Stack Sullivan, *The Interpersonal Theory of Psychiatry* (New York: Norton, 1953).

24. Robert D. Stolorow, Bernard Brandchaft, and George E. Atwood, *Psychoanalytic Treatment: An Intersubjective Approach* (Hillsdale, NJ: Analytic Press, 1987), ix.

25. Ronald S. Valle, Mark King, and Steen Halling, "An Introduction to Existential-Phenomenological Thought in Psychology," in *Existential-Phenomenological Perspectives in Psychology,"* ed. Ronald S. Valle and Steen Halling (New York: Plenum, 1989), 3–16.

26. Stolorow and Atwood, *Faces in a Cloud*, 20.

27. George Frank, "The Intersubjective School of Psychoanalysis: Concerns and Questions," *Psychoanalytic Psychology* 15, no. 3 (1998): 420–423.

28. Robert D. Stolorow, "Clarifying the Intersubjective Pesrpective: A Reply to George Frank," *Psychoanalytic Psychology* 15, no. 3 (1998): 424–427.

29. George E. Atwood and Robert D. Stolorow, "Experience and Conduct," *Contemporary Psychoanalysis* 17, no. 2 (1981): 197–208.

30. Irvin D. Yalom and Ginny Elkin, *Every Day gets a little Closer: A Twice-told Therapy* (New York: Basic Books, 1974).

31. See, for example, Carl R. Rogers, *Client-centered Therapy: Its Current Practice, Implications, and Theory* (Boston: Houghton Mifflin, 1965).

32. Stolorow, Atwood, and Orange, *Worlds of Experience*, 85.

33. Robert D. Stolorow, "Dynamic, Dyadic Intersubjective Systems: An Evolving Paradigm for Psychoanalysis," *Psychoanalytic Psychology* 14, no. 3 (1997): 337–346.

34. Esther Thelen and Linda B. Smith, *A Dynamic Systems Approach to the Development of Cognition* (Cambridge, MA: MIT Press, 1994).

35. Stolorow, "Dynamic," 337.

36. Stolorow, Atwood, and Orange, *Worlds of Experience*, 87.

37. Stolorow, "Dynamic," 339.

38. Stephen A. Mitchell, "Juggling Paradoxes: Commentary on the Work of Jessica Benjamin," *Studies in Gender and Sexuality* 1, no. 3 (2000): 251–269.

39. Judith Guss Teicholz, *Kohut, Loewald, and the Postmoderns: A Comparative Study of Self and Relationship* (Hillsdale, NJ: Analytic Press, 1999), 182–183.

40. Jessica Benjamin, *Like Subjects, Love Objects* (New Haven, CT: Yale University Press, 1995), 30.

41. Ibid., 29.

42. Stolorow, Atwood, and Orange, *Worlds of Experience*, 85–86.

43. Martin Buber, "Distance and Relation," in *The Knowledge of Man: A Philosophy of the Interhuman*, ed. Maurice Friedman (New York: Harper Torchbooks, 1966), 67.

44. Kirk J. Schneider, *Rediscovery of Awe: Splendor, Mystery, and the Fluid Center of Life* (St. Paul, MN: Paragon House, 2004); Karl Rahner, *The Love of Jesus and the Love of Neighbor* (New York: Crossroads, 1983); Sharon Salzberg, *Faith: Trusting Your Own Deepest Experience* (New York: Riverhead Books, 2003); Martin Buber, *I and Thou*, 2nd ed., trans. Ronald Gregor Smith (New York: Scribner and Sons, 1958).

45. Heather Menzies, *Fast Forward and Out of Control: How Technology is Changing Your Life* (Toronto: Macmillan of Canada, 1989; *No Time: Stress and the Crisis of Modern Life* (Vancouver: Douglas & McIntyre, 2005).

46. Thomas de Zengotita, "The Numbing of the American Mind: Culture as Anesthetic," *Harper's Magazine*, April 2002, 35.

47. Thomas Harpur, *Harpur's Heaven and Hell* (Toronto: Oxford University Press, 1983), 178.

Name Index

Aanstoos, Christopher, 230n33
American Psychiatric Association, 111,
    133, 134, 135, 226n34, 227n46
Andreasen, Nancy, 130–131, 141,
    226n32, 228n65
Anthony, William, 138
Antonovsky, A. M., 220n9
Arendt, Hannah, 83, 222n6
Arieti, Silvano, 69, 124–126, 222n43,
    226n23
Armstrong, Karen, 201, 235n2
Aron, Arthur, 202, 204, 235n5,
    236nn15–16
Aron, Elaine N., 236n12
Aron, Lewis, 206, 236n18
Atwood, George E., 122–124,
    206–211, 212, 226n21,
    236nn19–21, 236n24, 236n26,
    237n29, 237n32, 237n36,
    237n42

Barker, Elliott T., 114, 115, 225n11
Barlow, David H., 228n54
Baron, Robert A., 235n8
Barrett, William, 103, 137–138 139,
    164, 224n38, 228n57
Bauer, Lin, 220n2, 222n44, 223n7
Baum, Gregory, 79, 222n48
Baxter, L. A., 16, 218n4
Beck, Aaron, 102
Beck, Brittney, 231n42
Becker, Ernest, 186–192, 234nn17–20,
    234n30

Bemporad, Jules, 69, 222n43
Benjamin, Jessica, 211–212, 237n38,
    237n40
Berenson, Paulus, 77, 222n46
Berg, Michael, 67, 222n38
Berman, Harry L., 233n71
Bevan, William, 140–141, 143,
    228n67, 229n2
Binswanger, Ludwig, 148
Black, Margaret J., 61, 221n13
Blumenfeld, Laura, 81, 222n2
Boss, Medard, 148
Brandchaft, Bernard, 236n24
Briod, Marc, 232n66
Brown, G. W., 224n5
Buber, Martin, 25, 30, 93, 107, 114,
    206, 212, 213, 219nn15–16,
    219nn25–26, 219n28, 224n1,
    225n11, 237n43
Bullis, C., 16, 218n4
Bunney, William B., Jr., 127, 130,
    226n26
Burke, J. D., 225n9
Bush, George W., 60, 72
Byrne, Donne, 235n8

Carson Daniels, Christen, 67–68, 75,
    222n40
Carter, Jimmy, 110
Chambles, Diane, 228n55
Chesterton, G. K., 4–9, 113, 217nn8–9
Clingaman, Daniel, 232n54
Coenen, Herman, 233n70

Subject Index